Lecture Notes in Control and Information Sciences

Edited by M. Thoma

Lecture Notes in Control and Information Sciences

Edited by M. Thoma and A. Wyner

76

Stochastic Programming

Edited by
F. Archetti, G. Di Pillo and M. Lucertini

Springer-Verlag Berlin Heidelberg GmbH

Editors
Prof. F. Archetti
Dipartimento di Matematica
Università di Milano
Via L. Cicognara, 7
20129 Milano, Italy

Prof. G. Di Pillo
Prof. M. Lucertini
Dipartimento di Informatica
e Sistemistica
Università di Roma »La Sapienza«
Via Eudossiana, 18
00184 Roma, Italy

ISBN 978-3-540-16044-1   ISBN 978-3-540-39729-8 (eBook)
DOI 10.1007/978-3-540-39729-8

Library of Congress Cataloging in Publication Data

Main entry under title:
Stochastic programming.
(Lecture notes in control and information sciences; 76)
Selection of papers based on the contributions
discussed at the Working Conference on Stochastic
Programming held in Gargnano (Italy), September 15-21,
1983.
1. Stochastic programming--Congresses.
I. Archetti, Fancesco
II. Di Pillo, G.
III. Lucertini, M. (Mario)
IV. Working Conference on Stochastic Programming
(1983 : Gargnano, Italy)
V. Series.
T57.79.S75 1986   001.4'34   85-27845

2161/3020-543210

PREFACE

This volume contains a selection of papers based on the
contributions discussed at the Working Conference on Stochastic
Programming held in Gargnano (Italy), September 15-21, 1983.
The Conference was sponsored by the IFIP Technical Committee
on System Modelling and Optimization (TC-7), and organized with
the support of the University of Milan, which offered the
beautiful Villa Feltrinelli where the Conference took place,
and the National Council of Researches (C.N.R.) through the
Institute of Applied Mathematics and Informatics and the Institute of System Analysis and Informatics. Additional financial
aids were provided by the C.N.R. through the National Technical
Committees on Engineering Sciences, on Mathematical Sciences
and on Technological Sciences.

The Working Conference was attended by 48 participants from
16 countries, and 32 papers were presented and discussed.

All the contributed papers included in this volume have been
refereed and revised: the editors wish to thank for their
contribution in the refereeing process:
- A. Bertoni - N. Bellomo - D. Bertsekas - M. Dempster -
Y. Ermolev - E. Fagiuoli - P. Kall - D. Iglehart - L. Moore -
A. Prekopa - S. Provan - R. Wets - .

The papers collected here can be ranged in two main research
areas : stochastic modelling and simulation and stochastic
optimization . Both areas are of significant theoretical interest
and have a wide impact on to-day applications, as for instance
in flexible manufacturing systems, in computer networks, in
economic decision making and so on.

The International Program Committee of the working conference
consisted of A.V. Balakrishnan (University of California at
Los Angeles, U.S.A.), chairman, A. Bensoussan (I.N.R.I.A., France),
M. Cugiani (University of Milan, Italy), P. Kall (University of
Zurich, Switzerland), A. Ruberti (University of Rome "La Sapienza"
Italy) and of J. Stoer (University of Wurzburg, F.R.G.), which
we aknowledge for their advices in focusing the conference main
topics and in inviting the related speakers.

The conference secretary was Mrs. Anna Russo, which we
aknowledge for her kind assistance.

F. Archetti
G. Di Pillo
M. Lucertini

CONTENTS

V

MINIMAL TIME DETECTION OF PARAMETER CHANGE
IN A COUNTING PROCESS[*]

A. V. Balakrishnan
System Science Department
School of Engineering
UCLA
Los Angeles, CA 90025

Abstract

We present an algorithm for on-line detection of parameter change in a counting process (such as change in arrival rate in a queue), the optimality criterion being the
minimization of the time delay in detection. The development is based on the theory
of optimal stopping rules. An illustrative simulation study of a simple change model
is included.

1. Introduction

Suppose the arrival rate in a queue changes suddenly at some random time. How quickly can we detect this change? Is there an algorithm that is optimal in terms of
minimizing the (average) time delay of detection, keeping a given false alarm rate?
This is the problem we shall study in this paper.

Of course such a problem can occur in a wide variety of application areas -- in fact
in any situation where model parameters are subject to abrupt change at some random
time and we need to detect this change from whatever measurement data is available.
Hence we begin with a precise mathematical formulation of the general problem.

Let θ be a Markov time. We shall only consider discrete-time models, so that θ
assumes positive integral values n, $n \geq 1$. We do allow a nonzero probability for

$$\text{Pr.} \ (\theta = +\infty) \ = \ p_o \ \neq \ 0$$

$$= \ \text{Pr.} \ (\theta \neq n \ \text{ for any } \ n) \ \ .$$

Let $\{v_i\}$, $i = 1, 2, \ldots, n, \ldots$ denote measured (observed) data and let $\mathcal{F}_n$
denote the sigma algebra generated by $v_1, \ldots, v_n$. The problem is to find a Markov
time τ adapted to the growing sigma algebra $\mathcal{F}_n$ (in other words

$$(\tau = n) \ \epsilon \ \mathcal{F}_n \ )$$

* Research supported in part under Grant No. 78-3550, AFOSR, USAF, Applied Math
  Division

which minimizes the "time delay":

$$E[(\tau - \theta)^+]$$

subject to the condition that the "false alarm" probability

$$Pr. \ (\tau < \theta) \ \leq \ \alpha$$

for given α.

Such a problem was first considered by Kolmogorov -- the so-called "disruption" problem -- and subsequently elaborated by Shiryayev, see [1]. However Shiryayev considers a special case in which he can exploit the theory of Markov processes (and excessive functions), and it turns out that the assumptions he makes do not hold in applications of interest. In fact, and this is our purpose here, it is possible to obtain more general results (we are only concerned with discrete-time models) by going back to the basic theory of optimal stopping rules as developed by Chow, Robbins and Siegmund in [2].

2. General Theory

We consider only discrete-time models. We are given a Markov time θ taking on positive integral values and

$$Pr. \ (\theta = +\infty) \ = \ P_o$$

not necessarily zero, and an increasing sigma-algebra $\mathcal{F}_n$, $n \geq 1$. Let τ be a Markov time adapted to $\{\mathcal{F}_n\}$. Let us first consider the problem of minimizing

$$1 - E[\Pi_\tau] + cE[(\tau-\theta)^+] \tag{2.1}$$

where Π_n is the conditional probability

$$\Pi_n \ = \ Pr. \ (\theta \leq n \mid \mathcal{F}_n) \ . \tag{2.2}$$

We note that the first term in (2.1) is the probability of false alarm:

$$Pr. \ (\tau < \theta) \ = \ 1 - Pr. \ (\theta \leq \tau)$$
$$= \ 1 - E[\Pi_\tau] \ .$$

The first step in dealing with (2.1) is to cast in the form

$$E[x_\tau]$$

where x_n is adapted $\mathcal{F}_n$. Since the first term is already in this form we only need to work with the second term. Here we follow Shiryayev [1] who shows that

$$E[(\tau-\theta)^+] = \left[E \sum_1^{\tau-1} \Pi_k\right] .$$

Hence minimizing (2.1) is equivalent to maximizing

$$E[x_\tau] \qquad (2.3)$$

where

$$x_n = -\left(1 - \Pi_n + c \sum_1^{n-1} \Pi_k\right) . \qquad (2.4)$$

Now we follow [2] and invoke "backward induction." Thus we consider the special case where we restrict τ so that

$$\tau \leq N .$$

Then the optimal τ in this class is defined by

$$\text{Inf. } x_k = \gamma_k^N$$

where γ_k^N is defined iteratively as follows:

$$\gamma_N^N = x_N$$

$$\gamma_{N-1}^N = \text{Max } (x_N, E[\gamma_N^N \mid \mathcal{F}_{N-1}])$$

and generally

$$\gamma_m^N = \text{Max } (x_m, E[\gamma_{m+1}^N \mid \mathcal{F}_m]) , \qquad m \leq N-1 . \qquad (2.5)$$

The main point in introducing this step is that in our case we can calculate $\{\gamma_n^N\}$ as follows. For this purpose we need to calculate:

$$E[x_n \mid \mathcal{F}_{n-1}] .$$

For this in turn we need

$$E[\Pi_n \mid \mathcal{F}_{n-1}]$$

which can be calculated as follows:

$$E[\Pi_n \mid \mathcal{F}_{n-1}] = \text{Pr. } (\theta \leq n \mid \mathcal{F}_{n-1})$$

$$= \text{Pr. } (\theta \leq n-1 \mid \mathcal{F}_{n-1}) + \text{Pr. } (\theta = n \mid \mathcal{F}_{n-1})$$

$$= \Pi_{n-1} + \frac{(1-\Pi_{n-1}) \text{ Pr.}(\theta=n)}{\text{Pr. } (\theta < n-1)} . \qquad (2.6)$$

Let

$$a_{n-1} = \frac{\text{Pr. } (\theta=n)}{\text{Pr. } (\theta>n-1)} \quad . \tag{2.7}$$

Then

$$E[\Pi_n \mid \mathcal{F}_{n-1}] = a_{n-1} + (1-a_{n-1})\Pi_{n-1} \quad .$$

Hence

$$E[x_n \mid \mathcal{F}_{n-1}] = -1 + a_{n-1} + (1-a_{n-1})\Pi_{n-1} - c\Pi_{n-1} - c\sum_0^{n-2} \Pi_k$$

$$= x_{n-1} + g_{n-1}$$

where

$$g_k = a_k - (c+a_k)\Pi_k \quad . \tag{2.8}$$

Hence

$$\gamma_{N-1}^N = \text{Max } (x_{N-1}, \; x_{N-1}+g_{N-1})$$

$$= x_{N-1} + g_{N-1}^+$$

$$\gamma_{N-2}^N = \text{Max } [x_{N-2}, \; x_{N-2}+g_{N-2}+E[g_{N-1}^+\mid\mathcal{F}_{N-2}]]$$

$$= x_{N-2} + h_{N-2}^N$$

where

$$h_{N-2}^N = (\phi_{N-2}^N)^+$$

$$\phi_{N-2}^N = g_{N-2} + E[h_{N-1}^N\mid\mathcal{F}_{N-2}]$$

$$h_{N-1}^N = g_{N-1}^+ \quad .$$

More generally

$$\gamma_m^N = x_m + (g_m+\lambda_m^N)^+ \qquad m \le N-1$$

$$\lambda_m^N = E[h_{m+1}^N\mid\mathcal{F}_m]$$

$$h_m^N = (\phi_m^N)^+$$

$$\phi_m^N = g_m + E[h_{m+1}^N\mid\mathcal{F}_m]$$

$$\phi_{N-1}^N = g_{N-1} \quad .$$

Now we fix m and let $N \to \infty$. Then we know that γ_m^N converges. Let

$$\lim_{N} \gamma_m^N = \gamma_m \quad .$$

Then we have

$$\gamma_n = x_n + (g_n + \lambda_n)^+ \quad . \tag{2.9}$$

Now

$$\lambda_m^N \leq \sum_{k=1}^{N-m-1} E[g_{m+k}^+ | \mathcal{F}_m]$$

and hence

$$\lambda_m \leq \sum_{1}^{\infty} E[g_{m+k}^+ | \mathcal{F}_m] \tag{2.10}$$

and the right side is a super martingale which converges to zero. The appropriate "regularity" conditions being satisfied, it follows from [2] that the optimal stopping time τ is defined by

$$\tau = \inf_{n} [x_n = \gamma_n]$$

or, equivalently

$$\tau = \inf_{n} (g_n + \lambda_n)^+ = 0$$

or, equivalently, using (2.9):

$$\tau = \inf_{n} \Pi_n \geq \frac{a_n}{c + a_n} + \frac{\lambda_n}{c + a_n} \quad . \tag{2.11}$$

Note that the detection "threshold" is "time-varying." We can obtain a non-time-varying threshold (nonoptimal, of course) by setting $\lambda_n = 0$ (since $_n$ converges to zero) and taking

$$\tau = \inf_{n} \Pi_n \geq t$$

where

$$t = \inf_{n} \frac{a_n}{a_n + c} \quad .$$

There is one case where

$$\frac{a_n}{a_n + c} = \frac{p}{p + c} \tag{2.12}$$

and is independent of n, when the distribution of θ is geometric:

$$P_k = (1 - p_o)p(1-p)^{k-1} \quad , \qquad k \geq 1 \quad . \tag{2.13}$$

Another special case of interest is when

$$1 \leq \theta \leq N < \infty$$

and the distribution is uniform so that

$$P_k = (1-p_0)\frac{1}{N} \, , \qquad 1 \leq k \leq N \, .$$

In this case

$$\text{Inf.} \frac{a_k}{a_k + c} = \left(\frac{1}{N}\right) \frac{1}{\left(\frac{1}{N} + c\right)} \, ,$$

or, corresponds to taking

$$p = \frac{1}{N} \, .$$

Of course in this case

$$\tau \leq N \, .$$

Fixed False Alarm. It is possible to show following the arguments in Shiryayev [1] that minimizing

$$E[(\tau-\theta)^+]$$

subject to the fixed false alarm probability α:

$$\text{Pr.} \, (\tau<\theta) \leq \alpha$$

is equivalent to minimizing (2.1) for some c. However finding such a c for given α is apparently too difficult (see the discussion in Shiryayev [1]). Note that whatever the value of c, we do have that (2.1) for the optimal τ is

$$\leq 1 - E[\Pi_1] \tag{2.14}$$

since

$$E[(1-\theta)^+] = 0 \, .$$

It follows from (2.14) that

$$\alpha = 1 - E[\Pi_\tau]$$

$$\leq 1 - E[\Pi_1]$$

$$= 1 - (1-p_0)p$$

which puts an upper bound on α. For any fixed threshold t:

$$t = \underset{n}{\text{Inf.}} \, \Pi_n \geq t \tag{2.15}$$

we have that

$$= 1 - E[\Pi_\tau]$$

$$\leq 1 - t$$

or,

$$t \geq 1 - \alpha$$

but t may be strictly larger. Of course t must be determined from the stipulated false alarm. Also, (2.15) is not strictly optimal but close enough!

3. Application

Let us consider the following specific problem involving counting processes as an example. Let v_i, $i = 1, 2, \ldots,$ denote inter-arrival times in an M-G-1 queue. We shall consider only the discrete-time case so that v_i are integers, with

$$\text{Pr. } (v_i = n) = (1 - q_o) q_o^{n-1} \, , \qquad n \geq 1, \ldots \, . \qquad (3.1)$$

At some random time θ, the arrival rate changes suddenly to another value. Thus for $i > \theta$,

$$\text{Pr. } (v_i = n) = (1 - q_1) q_1^{n-1} \, . \qquad (3.2)$$

One may also use the more general change model where the change is randomized so that for $i > \theta$,

$$\text{Pr. } (v_i = n) = \sum_{i=1}^{N} \frac{1}{N} (1 - q_i) q_i^{n-1} \qquad (3.3)$$

for arbitrary N. The probability for no change is P_o:

$$\text{Pr. } (\theta \neq k \text{ for any } k) = P_o \, ,$$

and

$$\text{Pr. } (\theta = k) = P_k \, , \qquad k \geq 1 \, .$$

The main calculation involves Π_n. We have the general formula:

$$\Pi_n = \frac{\sum_1^n P_k(v_1, \ldots, v_n) P_k}{\sum_1^n P_k(v_1, \ldots, v_n) P_k + \left(1 - \sum_1^n P_k\right) P_o(v_1, \ldots, v_n)} \qquad (3.4)$$

where

$$P_k(v_1, \ldots, v_n) = p(v_1, \ldots, v_n \mid \theta = k)$$

$$P_o(v_1, \ldots, v_n) = p(v_1, \ldots, v_n \mid \theta > n) \, .$$

For our case

$$P_o(v_1,\ldots,v_n) = (1-q_o)^n \, q_o^{\sum\limits_1^n (v_i-1)} \tag{3.5}$$

$$P_k(v_1,\ldots,v_n) = (1-q_o)^{k-1} \, q_o^{\left(\sum\limits_1^{k-1}(v_i-1)\right)} P_1(v_k)\, P_1(v_{k+1}) \cdots P_1(v_n) \tag{3.6}$$

where $p_1(v)$ is specified by (3.2) or (3.3). Let

$$A_n = \sum\limits_1^n P_k(v_1,\ldots,v_n) p_k$$

$$B_n = \left(1 - \sum\limits_1^n p_k\right) P_o(v_1,\ldots,v_n) \quad .$$

Then the stopping rule can be expressed

$$\tau = \operatorname*{Inf.}_n \; \frac{A_n}{B_n} \geq t \quad . \tag{3.7}$$

Note that we can write:

$$\frac{A_n}{B_n} = \frac{1}{\left(1 - \sum\limits_1^n p_k\right)} \sum\limits_1^n (p_k)\frac{P_k(v_1,\ldots,v_n)}{P_o(v_1,\ldots,v_n)} \quad .$$

For the simple change model (3.2),

$$\frac{P_k(v_1,\ldots,v_n)}{P_o(v_1,\ldots,v_n)} = \left(\frac{1-q_1}{1-q_o}\right)^{n-k+1} \left(\frac{q_1}{q_o}\right)^{\sum\limits_k^n (v_i-1)} \quad . \tag{3.8}$$

In this problem, we note that Π_n is Markov, so that the assumptions of Shiryayev [1] hold if in addition we take the distribution of θ to be geometric (2.13). This would imply that there does exist a time-invariant threshold and that (3.7) is even more close to the optimal rule.

4. Simulation Results

In this section we present some results of a simple simulation study in which the nonrandom change model (3.2) is used, along with (2.13). The basic purpose of the simulation is to examine the behavior of the decision function:

$$\operatorname{Log}_{10} \frac{A_n}{B_n} \quad ,$$

and is plotted in Figures 1 through 6 for various values of q_o, q_1 and p, with

$$P_o = 1$$

in all cases. As we have noted, the threshold for a given false alarm depends on the parameter chosen. All the plots show a sharp break at the instant of change and we see that the dependence on the Bayesian parameters is not critical.

References

[1] A. N. Shiryayev: Optimal Stopping Rules. Springer-Verlag, New York, 1978.

[2] Y. S. Chow, H. Robbins and D. Siegmund: Great Expectations: The Theory of Optimal Stopping. Houghton Mifflin Co., New York, 1971.

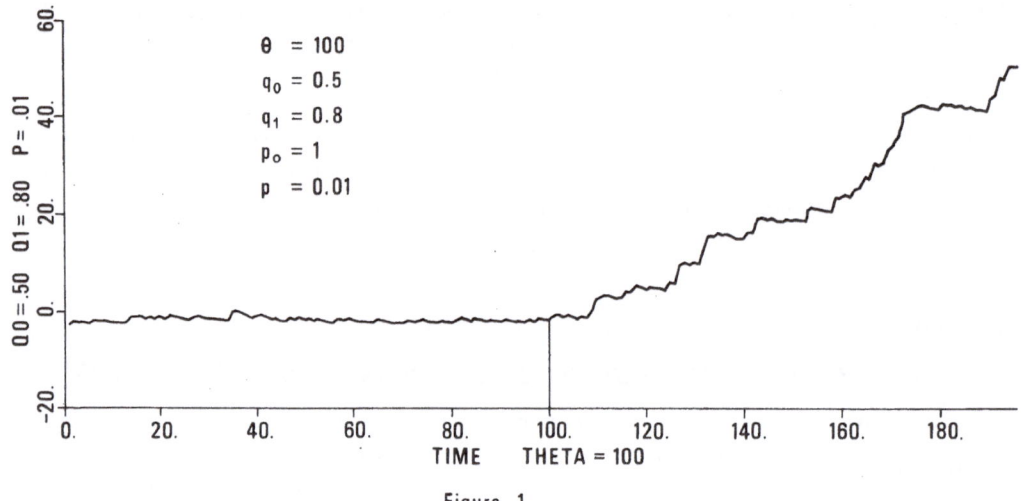

Figure 1

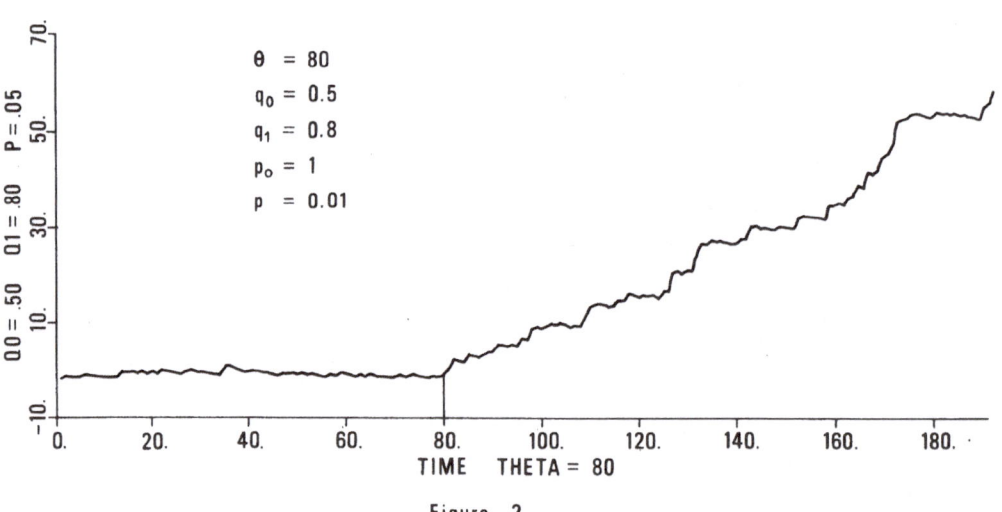

Figure 2

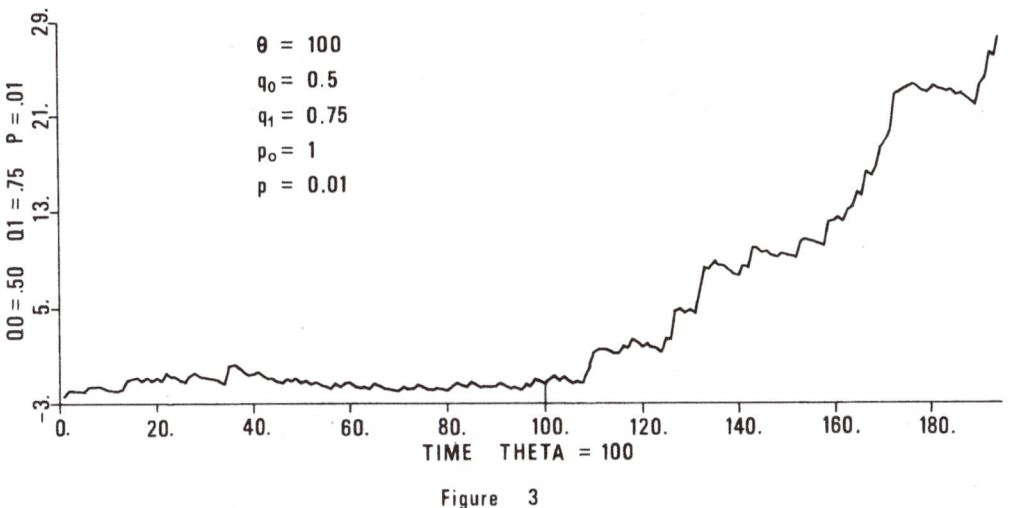

Figure   3

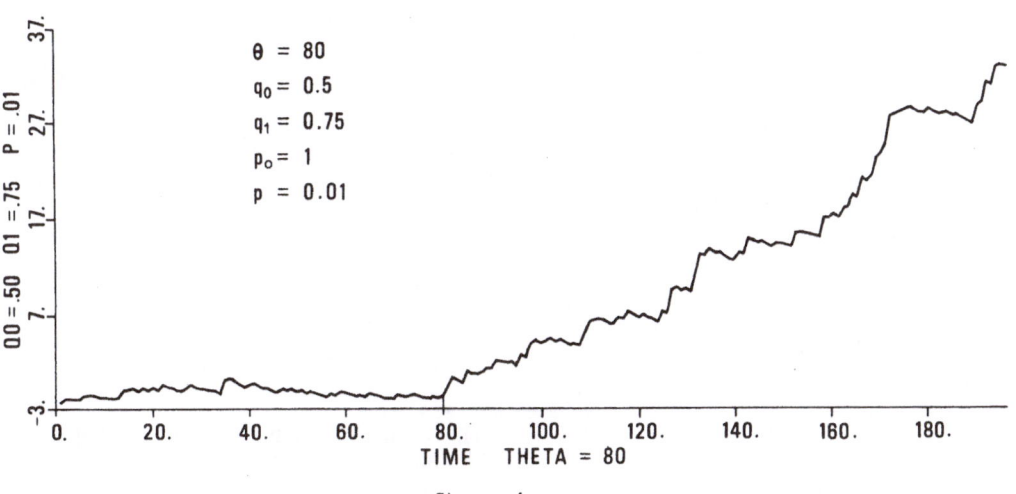

Figure   4

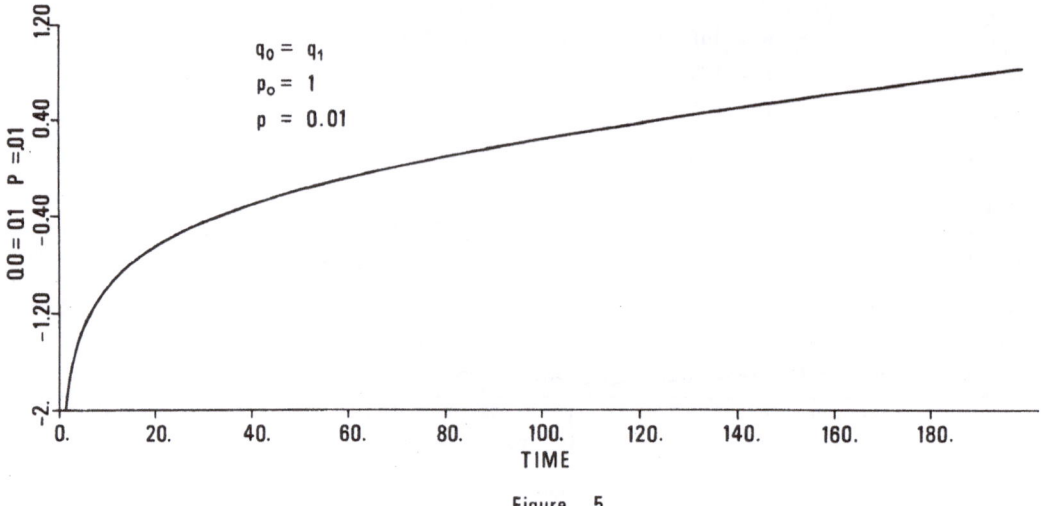

Figure 5

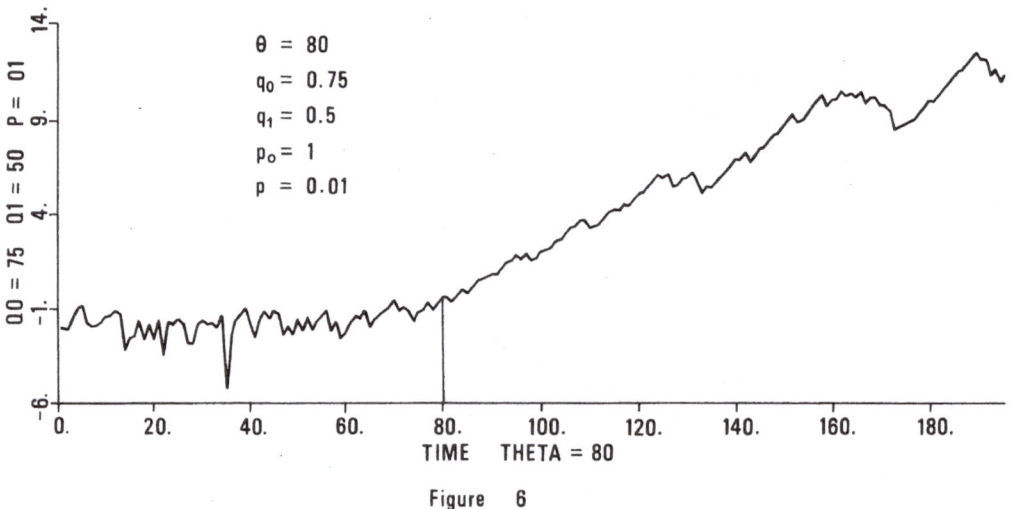

Figure 6

SIMULATION FOR PASSAGE TIMES IN NON-MARKOVIAN NETWORKS OF QUEUES

Donald L. Iglehart

Department of Operations Research

Stanford University

Stanford, California 94305

Gerarld S. Shedler

IBM Research Laboratory

San Jose, California 95193

ABSTRACT

   An appropriate state vector for simulation of closed networks of queues
with priorities among job classes is a linear "job stack", an enumeration of
service center and job class of all the jobs. Simulation for passage times
can be based on observation of an augmented job stack process which maintains
the position of an arbitrarily chosen "marked job". Using a representation
of the augmented job stack process as a generalized semi-Markov process, we
develop an estimation procedure for passage times in networks with general
service times. We also describe an estimation procedure for passage times
which correspond to the passage through a subnetwork of a given network of
queues. With this "labelled jobs method", observed passage times for all the
jobs are used to construct point and interval estimates. Our results apply
to networks with "single states" for passage times. Based on a single simulation run, the procedures provide point estimates and confidence intervals
for characteristics of limiting passage times.

1. INTRODUCTION

Assessing the statistical precision of a point estimate obtained from the simulation of a stochastic system requires careful design of the simulation experiments and analysis of the simulation output. In general, the desired statistical precision takes the form of a confidence interval for the quantity of interest. Among the issues the simulation analyst must face are the initial conditions for the system being simulated, the length of the simulation run, the number of replications of the experiments, and the length of the confidence interval. Based on limit theorems for regenerative stochastic processes, the regenerative method (Crane and Iglehart [2]) is a theory of simulation analysis which, when applicable, provides some measure of statistical precision; see Crane and Lemoine [3] for an introduction to regenerative simulation.

The regenerative method is the basis for previous work (Iglehart and Shedler [8]) on simulation methods for "passage times" in networks of queues with priorities among job classes. Passage times (informally, the random times for a job to traverse a portion of a network) are important in connection with computer and communication system models, and in this context, expected values as well as other characteristics of passage times are of interest.

The estimation procedures for passage times developed in [8] are applicable to networks with priorities among job classes in which all service time distributions have a Cox-phase (exponential stage) representation (Cox [1]); this assumption preserves the Markovian structure of the model. In this paper, we describe two regenerative estimation procedures which avoid Cox-phase representation of general service time distributions. Our estimation procedures are applicable to networks that have a "single state" for passage times. Regenerative cycles are defined in terms of the single state. The *marked job method* prescribes observation of passage times for an arbitrarily chosen, distinguished job. The *labelled jobs method* (Shedler and Southard [20]) provides estimates for particular passage times which correspond to passage through a subnetwork of a given network of queues. With the labelled jobs method, observed passage times for all the jobs are used to construct point and interval estimates.

An appropriate state vector for closed, multiclass networks of queues with priorities among job classes is a linear "job stack", an enumeration by service center and job class of all the jobs. Simulation for passage times can be based on the observation of an augmented job stack process which maintains the position of an arbitrarily chosen "marked job". Using a re-presentation as an irreducible generalized semi-Markov process, we show that the augmented job stack process restricted to an appropriate subset of its state space is a regenerative process in continuous time. The choice of a particular sequence of regeneration points leads to an estimation procedure for passage times which is based on observation of one sample path of the restricted augmented job stack process. This marked job method provides strongly consistent point estimates and asymptotic confidence intervals for general characteristics of limiting passage times.

For the labelled jobs method passage times for all the jobs are recorded by observing a "fully augmented job stack process", which maintains the position of each of the jobs in the job stack. Under a mild restriction on the priorities among job classes, the job stack process observed at the epochs at which passage times terminate is a regenerative process in discrete time. As a consequence, point and interval estimates for characteristics of limiting passage times can be obtained from a single simulation run. Termi-nations of passage times with no other passage times underway and exactly one job in service are regeneration points for the job stack process observed at termination times. In order for such epochs to exist we must exclude passage times which always terminate with two or more jobs in service. A mild restriction on the priorities among job classes ensures that infinitely many such epochs occur.

It may be possible to develop valid estimation procedures for passage times based on spectral methods (cf. Heidelberger and Welch [7]), the method of batch means (cf. Fishman [4], Law and Carson [14]), or the method of independent replicates (cf. Fishman [4]). However, the validity of these

methods as general techniques for simulation output analysis rest on particular assumptions which are not necessarily satisfied by an output sequence
of passage times. For example, spectral methods assume that the output
sequence is second-order stationary, and the observed sequence of passage
times is asymptotically stationary but not second-order stationary. Valid
estimation procedures for passage times based on these methods are of interest.

2. CLOSED, MULTICLASS NETWORKS OF QUEUES AND PASSAGE TIMES

   As in $\begin{bmatrix} 8 \end{bmatrix}$, we consider closed networks of queues having a finite number
of *jobs* (customers), N, a finite number of *service centers*, s, and a finite
number of (mutually exclusive) *job classes*, c. At every epoch of continuous
time each job is in exactly one job class, but jobs may change class as
they traverse the network. Upon completion of service at center i a job of
class j goes to center k and changes to class l with probability $p_{ij,kl}$,
where

$$P = \{ p_{ij,kl} : (i,j),(k,l) \in C \}$$

is a given irreducible stochastic matrix and $C \subseteq \{1,2,\ldots,s\} \times \{1,2,\ldots,c\}$
is the set of (center, class) pairs in the network. At each service center
jobs queue and receive service according to a fixed priority scheme among
classes; the priority scheme may differ from center to center. Within a class
at a center, jobs receive service according to a fixed queue service discipline; e.g., first-come, first-served (FCFS). Note that in accordance with
the matrix P, some centers may never see jobs of certain classes. According
to a fixed procedure for each center, a job in service may or may not be
pre-empted if another job of higher priority joins the queue at the center.
(The interruption of service is assumed to be of the preemptive-repeat type).
A job that has been preempted samples a new service time from the appropriate
distribution and receives this additional service at the center before any
other job of its class at the center receives service.

   All service times are assumed to be mutually independent. We also

suppose that service times at a center have finite mean but otherwise
arbitrary density function which is continuous and positive on $(0,\infty)$. Parameters of the service time distribution may depend on the service center,
the class of job being served and the "state" (as defined in Equation (2.1)
below) of the entire network. In order to characterize the state of the
network at time t, we let $S_i(t)$ denote the class of the job receiving service
at center i at time t, where $i = 1,2,\ldots s$; by convention $S_i(t)=0$ if at time
t there are no jobs at center i. If center i has more than one server, we
enumerate the servers at center i and let $S_i(t)$ be a vector which records
the class of the job receiving service from each server at the center. (A
job receives service from the lowest numbered available server). The classes
of jobs serviced at center i ordered by decreasing priority are

$j_1(i),j_2()i),\ldots j_{k(i)}(i)$, elements of the set $\{1,2,\ldots,c\}$. We denote by
$c_{j_1}^{(i)}(t),\ldots c_{j_{k(i)}}^{(i)}(t)$ the number of jobs in queue at time t of the various
classes of jobs serviced at center i, $i = 1,2,\ldots,s$.

We order the N jobs in a linear stack (column vector) according to the
following scheme. For $t\geq 0$ define the state vector $Z(t)$ at time t by

$$Z(t) = (c_{j_{k(1)}}^{(1)}(t),\ldots,c_{j_1}^{(1)}(t),S_1(t);\ldots;c_{j_{k(s)}}^{(s)}(t),\ldots c_{j_1}^{(s)}(t),S_s(t)). \quad (2.1)$$

The *job stack at time t* then corresponds to the nonzero components of the
vector $Z(t)$ and thus orders the jobs by class at the individual centers.
Within a class at a particular service center, jobs waiting appear in the job
stack in FCFS order; i.e., jobs appear in order of their arrival at the
center, the latest to arrive being closest to the top of the stack. The
process $Z = \{Z(t):t\geq\}$ is called the *job stack process*. For any service
center i that sees only one job class (i.e., such that $k(i) = 1$), it is
possible to simplify the state vector by replacing $c_{j_{k(i)}}^{(i)}(t)$, $S_i(t)$ by $Q_i(t)$,
the total number of jobs at center i. Note that the state vector definition
does not take into account explicitly that the total number of jobs in the
network is fixed. In the case of complex networks, the use of this resulting
somewhat larger state space facilitates generation of the state vector

process; for relatively simple networks, it may be desirable to remove the redundancy.

Definition of passage times

Denote by $N(t)$ the position (from the top) of the marked job in the job stack at time t. (For example, the marked job is at the head of the line of the class $j_{k(1)}$ queue at center 1 if $c_{j_{k(1)}}^{(1)}(t) = n>0$ and $N(t) = n$.). Then set

$$X(t) = (Z(t),N(t)) \tag{2.2}$$

and call $X = \{X(t):t\geq0\}$ the *augmented job stack process*. Passage times are specified in terms of the marked job by means of four subsets $(A_1,A_2,B_1,$ and $B_2)$ of the state space, G^*, of the augmented job stack process X. The sets A_1,A_2 [resp. B_1,B_2] jointly define the random times at which passage times for the marked job start [resp. terminate]. The sets A_1,A_2, B_1 and B_2 in effect determine when to start and stop the clock measuring a particular passage time of the marked job.

Denote the jump times of the process X by $\{\tau_n:n\geq0\}$. For $k,n\geq1$ we require that the sets A_1, A_2, B_1, and B_2 satisfy the following conditions:

if $X(\tau_{n-1}) \in A_1$, $X(\tau_n)\in A_2$, $X(\tau_{n-1+k}) \in A_1$ and $X(\tau_{n+k}) \in A_2$

then $X(\tau_{n-1+m}) \in B_1$ and $X(\tau_{n+m}) \in B_2$ for some $0<m\leq k$;

and

if $X(\tau_{n-1}) \in B_1$, $X(\tau_n) \in B_2$, $X(\tau_{n-1+k}) \in B_1$ and $X(\tau_{n+k}) \in B_2$

then $X(\tau_{n-1+m}) \in A_1$ and $X(\tau_{n+m}) \in A_2$ for some $0\leq m<k$

These conditions ensure that the start and termination times for the specified passage time strictly alternate.

In terms of the sets A_1, A_2, B_1 and B_2, we define two sequences of random times, $\{S_j:j\geq0\}$ and $\{T_j:j\geq1\}$, where S_{j-1} is the start time of the jth passage time for the marked job and T_j is the termination time of this jth passage time. Assuming that the initial stage of the augmented job stack process X is such that a passage time for the marked job begins at $T = 0$, set

$$S_0 = 0$$

$$S_j = \inf\{\tau_n \geq T_j : X(\tau_n) \in A_2, \; X(\tau_{n-1}) \in A_1\}, \; j \geq 1$$

and

$$T_j = \inf\{\tau_n > S_{j-1} : X(\tau_n) \in B_2, \; X(\tau_{n-1}) \in B_1\}, \; j \geq 1.$$

Then the jth passage time for the marked job is $P_j = T_j - S_{j-1}$, $j \geq 1$. For passage times that are complete circuits in the network, $A_1 = B_1$ and $A_2 = B_2$; consequently $S_j = T_j$ for all $j \geq 1$.

For $z \in D^*$, the state space of Z, let $U(z)$ be the set of all $(i,j) \in C$ such that in state z there is a job of class j in service at center i. For $z, z' \in D^*$ and $u = (i,j) \in U(z)$, let $q(z';z,u)$ be the probability that the job stack process Z jumps (in one step) to state z', given that in state z there is a completion of service to a job of class j at center i. For all $z, z' \in D^*$, we say that z' *is accessible from z* and write $z \sim z'$ if there exists a finite sequence $u_0', z_1, u_1', \ldots, u_n'$ of (center, class) pairs and job stacks such that

$$q(z_1; z, u_0') q(z_2; z_1, u_1') \ldots q(z'; z_n, u_n') > 0 \qquad (2.3)$$

When $z \sim z'$ and $z' \sim z$ we say that *z and z' communicate* and write $z \approx z'$.

Analogously, we define $U(x)$ for $x \in G^*$: $U(x) = U(z)$ when $x = (z,n)$ for some $z \in D^*$ and $n \in \{1,2,\ldots,N\}$. For $x, x' \in G^*$ and $u = (i,j) \in U(x)$, we denote by $p(x'; x, u)$ the probability that the augmented job stack process X jumps to state x', given that in state x there is a completion of service to a job of class j at center i. We say that x' is *accessible from* x and write $x \sim x'$ if there exists a finite sequence $u_0', x_1, u_1', \ldots u_n'$ of (center, class) pairs and augmented job stacks such that

$$p(x_1; x, u_0') p(x_2; x_1, u_1') \ldots p(x'; x_n, u_n') > 0 \qquad (2.4)$$

The procedure given in Section 4 provides estimates for characteristics of limiting passage times for the marked job. In the absence of some restriction on the building blocks of a network of queues with priorities among job classes, the sequence of passage times for the marked job need not converge in distribution to a random variable independent of the initial state of the system. We make the further assumption that for some $z^* \in D^*$ the sets

$$D = \{z \in D^* : z^* \sim z\} \tag{2.5}$$

and

$$G = \{(z,n) \in G^* : z \in D\} \tag{2.6}$$

are irreducible in the sense that $z \sim z'$ for all $z, z' \in D$, and $x \sim x'$ for all $x, x' \in G$.

For networks with more than one service center ($s>1$), it is sufficient that for some service center, i_0, either $k(i_0) = 1$ or service at center i_0 to a job of class $j_{k(i_0)}(i_0)$ (the lowest priority job class seen by center i_0) is preempted when a job of higher priority joins the queue. Let $z^*_{i_0} \in D^*$ be the state in which there is one job of class $j_{k(i_0)}(i_0)$ in service at center i_0 and $N-1$ jobs of class $j_{k(i_0)}(i_0)$ in queue at center i_0 (or in service if center i_0 is a multiple server center). Define D to be the set of all states of the job stack process Z that are accessible from $z^*_{i_0}$ (i.e., $D = \{z \in D^* : z^*_{i_0} \sim z\}$) and set $G = \{(z,n) \in G^* : z \in D\}$.

(2.7) LEMMA. Suppose that the routing matrix P is irreducible and that for some service center i_0 either $k(i_0)=1$ or service to a job of class $j_{k(i_0)}(i_0)$ at center i_0 is preempted when a job of higher priority joins the queue. Then $z \sim z'$ for all $z, z' \in D$.

The idea (cf. Shedler and Slutz [19]) is to show that $z^*_{i_0}$ is a "target state" in the sense that it is accessible from any state in D. The proof is constructive and rests on the existence of a finite length "path" from $(i_0, j_{k(i_0)}(i_0))$ to $(i_0, j_{k(i_0)}(i_0))$ which includes all (center, class) pairs. Proposition (2.8) is established by showing (cf. [19]) that $x^*_{i_0} = (z^*_{i_0}, N)$ can serve as a target state; i.e., $x \sim x^*_{i_0}$ for all $x \in G$.

(2.8) PROPOSITION. Let the number of service centers s>1. Suppose that the routing matrix P is irreducible and that for some service center, i_0 either $k(i_0) = 1$ or service to a job of class $j_{k(i_0)}(i_0)$ at center i_0 is preempted when a job of higher priority joins the queue. Then $x \sim x'$ for all $x,x' \in G$.

Example (2.9) gives a multiclass representation of the network of queues defined by Lavenberg and Shedler [12] as a model of resource contention in the "DL/I component" of an IMS (Information Management System) data base management computer system.

(2.9) EXAMPLE. Consider a network (cf. [8], Section 5.2) with two service centers and seven job classes such that the set C of (center, class) pairs is $C = \{(1,2), (1,3), (1,4), (1,5), (1,6), (1,7), (2,1)\}$ and the classes of jobs serviced at center 1 ordered by decreasing priority are $j_1(1)=2$, $j_2(1)=3,\ldots,j_6(1)=7$. Service to jobs of class 7 (at center 1) is subject to preemption when a job of higher priority joins the queue at center 1. Service to any other job class is not subject to preemption. Also suppose that (for $0<\psi_1,\psi_2<1$) the routing matrix P is

	(1,2)	(1,3)	(1,4)	(1,5)	(1,6)	(1,7)	(2,1)
(1,2)	0	$1-\psi_1$	0	0	0	0	ψ_1
(1,3)	$1-\psi_2$	0	0	0	0	ψ_2	0
(1,4)	0	$1-\psi_1$	0	0	0	0	ψ_1
(1,5)	$1-\psi_2$	0	0	0	0	ψ_2	0
(1,6)	0	0	1	0	0	0	0
(1,7)	0	0	0	0	1	0	0
(2,1)	0	0	0	1	0	0	0

$P =$

Let

$$Z(t) = (c_7^{(1)}(t),\ldots,c_2^{(1)}(t),S_1(t),Q_2(t)),$$

where the number of jobs of class j in queue at center 1 at time t is $c_j^{(1)}(t)$, $S_1(t)$ is the class of the job in service at center 1, and $Q_2(t)$ is the number of jobs waiting or in service at center 2. As $k(2)=1$ and service to jobs of class $j_{k(1)}(1) = 7$ at center 1 is subject to preemption, either state $z_1^* = (N-1,0,0,0,0,0,7,0)$ or state $z_2^* = (0,0,0,0,0,0,0,N)$ can

serve as a target state for the job stack process. For $N \geq 2$ jobs the set

$D^*-D \neq \phi$ is nonempty,; e.g., the state $(0,0,0,k-1,0,0,4,N-k)$ is an element of

D^*-D provided that $k \geq 2$. Now let $N(t)$ denote the position of the marked job

in the job stack corresponding to the nonzero components of $Z(t)$, and set

$X(t) = (Z(t), N(t))$. Either state $x_1^* = (N-1,0,0,0,0,0,7,0,N)$ or state

$x_2^* = (0,0,0,0,0,0,0,N,N)$ can serve as a target state for the augmented job

stack process X.

Henceforth, we assume that the subsets A_1, A_2, B_1 and B_2 which define

the starts and terminations of passage times for the marked job are subsets

of G. Without loss of generality, we also suppose that $X(0) \in G$; thus, we

consider simulation of the augmented job stack process X restricted to the

set G.

Single states for passage times

Define a set S according to

$$S = \{(k,m): k \in A_1, m \in A_2 \text{ and } p(m;k,u) > 0 \text{ for some } u \in U(k)\} \qquad (2.10)$$

The entrances of the augmented job stack process to state m from state k

correspond to the starts of passage times for the marked job. We write

$h(z,n) = (i,j)$ when the job in position n in the job stack associated with

state $z \in D$ is of class j at center i, $n = 1,2,...,N$. Now define a subset

S' of S according to

$$S' = \{(z,N,z',n') \in S: \text{ for some single server center i and some}$$
$$(i,j_1(i)) \subset C,$$
$$h(z,N) = (i,j_1(i) \text{ and } h(z,n) = (i,j_{1_n}(i)) \text{ with } 1_n \geq 1, \ 1 \leq n < N\} \qquad (2.11)$$

and assume that $S' \neq \phi$. This condition ensures that there exists a state,

(z,N), of the augmented job stack process X such that the marked job is in

service at some single server center, the other $N-1$ jobs are in queue at

the center as jobs of equal or lower priority, and with positive probability

a passage time for the marked job starts upon completion of the service in progress. (In the context of generalized semi-Markov processes, these states are "single states" in the sense of Fossett [5]). These states are used to select return states in the procedures developed in the following sections.

3. THE UNDERLYING STOCHASTIC STRUCTURE

The estimation procedure for passage times given in Section 4 is based on the result that the augmented job stack process X defined by Equation (2.2) is a regenerative process in continuous time. To show this, we use a representation of the process (restricted to the set G) as an irreducible generalized semi-Markov process (Matthes [16], König, Matthes and Nawrotzki [11], [12]).

Following Whitt [22], a GSMP moves from state to state in accordance with the occurrence of events associated with the occupied state. Each of the several possible events associated with a state has its own jump distribution for determining the next state, and these events compete with respect to triggering the next transition. At each transition of the GSMP, new events may be scheduled. For each of these new events, a clock indicating the time until the event is scheduled to occur is set according to an independent (stochastic) mechanism. If a scheduled event does not trigger a transition but is associated with the next state, its clock continues to run; if such an event is not associated with the next state, it is abandoned.

The GSMP associated with the augmented job stack process has state space G and the events associated with state $x \in G$ are completions of service to a job of class j at center i, $(i,j) \in U(x)$. Jumps of the process are governed by the probability mass function $p(\cdot\,;x,u)$. At a jump from state x to state x' triggered by event u, new clock values are generated for each $v \in (x') - (U(x) - \{u\})$. The distribution function of such a new clock value (a service time for a job class at some center) is denoted by $F(\cdot\,;x',v,x,u)$; by our assumptions in Section 2, each has finite mean and a density function which is continuous and positive on $(0,\infty)$. For $v \in U(x') \cup (U(x) - \{u\})$, the old clock reading is kept after the jump. For $v \in (U(x) - \{u\}) - U(x')$, event v ceases to be scheduled after the jump. Let $U = \{u_1, u_2, \ldots, u_e\}$ be the set of all possible events that can occur; i.e.,

$$U = \bigcup_{\dot{x} \in G} U(x).$$

With each $x \in G$, associate the set of clock readings

$$C(x) = \{(c_1, c_2, \ldots c_e): c_i \geq 0 \text{ and } c_i > 0 \text{ if and only if } u_i \in U(x)\},$$

where c_i is the reading on the clock corresponding to event $u_i \in U(x)$. Note that the GSMP is irreducible since (by assumption) $x \sim x'$ for all $x, x' \in G$. (A GSMP is said to be irreducible if for all $x, x' \in G$ there exists a finite sequence $u_0', x_1, u_1', \ldots u_n'$ of events and states satisfying Equation (2.4).

(3.1) PROPOSITION. Let the number of service centers $s > 1$. Suppose that the routing matrix P is irreducible and for some service center, i_0, either $k(i_0) = 1$ or service to a job of class $J_{k(i_0)}(i_0)$ at center i_0 is preempted when a job of higher priority joins the queue. Also suppose that $S' \neq \phi$ and define

$$A'_2 = \{(z', n') \in A_2: (z, N, z', n') \in S' \text{ for some } z \in D\}. \tag{3.2}$$

Then $P\{X(S_n) = x' \text{ infinitely often}\} = 1$ for any $x' \in A'_2$.

Proof: We appeal to Glynn [6] to show that the general state space Markov chain (GSSMC) associated with state transitions of the GSMP returns infinitely often to the set $\{x'\} \times C(x')$; it then follows immediately that $P\{X(S_n) = x' \text{ infinitely often}\} = 1$. Three conditions must be checked:

(i) the GSMP is irreducible in the sense that $x \sim x'$ for all $x, x' \in G$;

(ii) the density functions associated with the clock readings c_i have finite means and are continuous and positive on $(0, \infty)$; and

(iii) a "recurrence measure" assigns positive measure to the set $\{x'\} \times C(x')$ for $x' \in A'_2$.

Condition (i) holds as a consequence of Proposition (2.8) and condition (ii) holds by assumption. With respect to condition (iii), note that the set

$C(x')$ is of the form

$$C(x') = \{(0,\ldots,0,c_i,0,\ldots,0,c_j,0,\ldots,0): c_i,c_j>0\} ,$$

where c_i and c_j are the readings on the two clocks active in state x'. One clock pertains to the marked job which is just starting a new service time. The other clock pertains to the job which is starting service at the center vacated by the marked job. The recurrence measure defined by Glynn assigns infinite measure to the set $\{x'\}\times C(x')$, $x'\in A_2'$; hence, condition (iii) holds and the GSSMC hits $\{x'\}\times C(x')$ infinitely often with probability one.

Now select an element x' from the set A_2' of Equation (3.2). From Proposition (3.1) we know that $\{X(S_n):n\geq 0\}$ hits x' infinitely often with probability one. Furthermore, at such a start time, S_n, the only (two) clocks that are active have just been set as described in the proof of Proposition (3.1). Since the jumps of the augmented job stack process X are governed by the Markovian mass function $p(\cdot;x,u)$ and the only active clocks have just been set at time S_n, the future evolution of the process X is independent of the history of the process before S_n and has the same distribution as it does when $X(0) = x'$. Thus, the subsequence of start times, S_n, at which $X(S_n) = x'$ are regeneration points for the process X. Since the state space of the augmented job stack process is finite and the clock setting distributions have finite mean, the expected time between regeneration points is finite.

(3.3) PROPOSITION. The process $\{X(t):t\geq 0\}$ is a regenerative process in continuous time and the expected time between regeneration points is finite.

From the argument leading to Proposition (3.3) it is clear that the random indices β_n such that $X(S_{\beta_n}) = x'$ constitute a sequence of regeneration points for the process $\{(X(S_n),P_{n+1}): n\geq 0\}$; this follows from the fact that the process $\{P_n:n\geq 1\}$ starts from scratch when $X(S_{\beta_n}) = x'$.

(3.4) PROPOSITION. The process $\{(X(S_n),P_{n+1}):n\geq 0\}$ is a regenerative process in discrete time and the expected time between regeneration points is finite.

The regenerative property guarantees (Miller [17]) that

$$(X(S_n), P_{n+1}) \implies (X, P) \tag{3.5}$$

as n→∞, i.e., there exist random variables X and P such that

$$\lim_{n \to \infty} P\{X(S_n) = i, P_{n+1} \leq y\} = P\{X=i, P \leq y\} \equiv F(i,y)$$

for all $i \in A_2$ and $y \in [0,\infty)$ for which F(i,·) is continuous.

The goal of the simulation is the estimation of $r(f) = E\{f(P)\}$, where f is a real-valued (measurable) function with domain (0,∞) and P is the limiting passage time for the marked job. It is intuitively clear and can be shown (cf. Appendix 1 of [8]) that the sequence of passage times for any other job (as well as the sequence of passage times, irrespective of job identity, in order of start or termination) converges in distribution to the same random variable as the sequence of passage times for the marked job.

4. SIMULATION FOR PASSAGE TIMES

Select x' ∈ A_2', begin the simulation of the augmented job stack process X with X(0) =x', and carry out the simulation of the process X in cycles defined by the successive entrances of $\{X(S_n):n \geq 0\}$ to x'. Let α_m denote the length (in discrete time units) of the mth cycle of $\{X(S_n):n \geq 0\}$ and define $\beta_0 = 0$ and $\beta_m = \alpha_1 + \ldots + \alpha_m$, m≥1; α_m is the number of passage times for the marked job in the mth cycle. Also define

$$Y_m(f) = \sum_{j=\beta_{m-1}+1}^{\beta_m} f(P_j).$$

Proposition (4.1) follows (cf. Crane and Iglehart [2]) from the fact that the random times $\{\beta_m:m \geq 0\}$ are regeneration points for $\{(X(S_n), P_{n+1}):n \geq 0\}$.

(4.1) PROPOSITION. The sequence of pairs of random variables $\{(Y_m(f), \alpha_m):m \geq 1\}$ are independent and identically distributed.

The final step is to establish a ratio formula for $E\{f(P)\}$. This follows from the general result for regenerative processes (cf. Crane and Iglehart [2], Proposition A.3).

(4.2) PROPOSITION. Let D(f) be the set of discontinuities for the function f. Provided that $P\{P \in D(f)\} = 0$ and $E\{|f(P)|\}<\infty$,

$$E\{f(P)\} = \frac{E\{Y_1(f)\}}{E\{\alpha_1\}} \; .$$

Given Propositions (4.1) and (4.2), the standard regenerative method ([3]) applies and (from a fixed number, n, of cycles) provides the strongly consistent point estimate

$$\hat{r}_n(f) = \frac{\bar{Y}_n(f)}{\bar{\alpha}_n}$$

for r(f), where

$$\bar{Y}_n(f) = \frac{1}{n} \sum_{m=1}^{n} Y_m(f)$$

and

$$\bar{\alpha}_n = \frac{1}{n} \sum_{m=1}^{n} \alpha_m \; .$$

Confidence intervals for r(f) are based on the central limit theorem

$$\frac{n^{1/2}\{\hat{r}_n(f) - r(f)\}}{\sigma(f)/E\{\alpha_1\}} \implies N(0,1)$$

as $n \to \infty$, where $\sigma^2(f)$ is the variance of $Y_1(f)-r(f)\alpha_1$ and N(0,1) is a standardized (mean 0, variance 1) normal random variable.

Application of the estimation procedure requires the selection of a return state $x' \in A_2' \subseteq G$. For complex networks it is nontrivial to determine the set G by inspection. Since $G = \{(z,n) \in G^*:z \in D\}$, it is sufficient to determine the elements of the set D defined by Equation (2.6). It is easy (cf. Proposition (4.1) of [19]) to characterize the elements of D.

(4.3) PROPOSITION. For i=1,2,...,s denote by $z_i^* \in D^*$ the state of the job stack process in which all N jobs are of class $j_{k(i)}(i)$ at center i. Assume that the routing matrix P is irreducible and that for some service center,

i_0, either $k(i_0)=1$ or service to a job of class $j_{k(i_0)}(i_0)$ at center i_0 is preempted when a job of higher priority joins the queue. Let $z \in D^*$. Then $z \in D$ if and only if $z_i^* \sim z$ for some service center $i, i=1,2,\ldots,s$.

(4.4) EXAMPLE. Consider a network with two service centers and two job classes such that the set C of (center, class) pairs is C= $\{(1,1),(2,2)\}$. Taking into account the fixed number of jobs in the network, let $Z(t)$ be the number of jobs waiting or in service at center 1 at time t. Also suppose that the irreducible routing matrix P is

$$
\begin{array}{ccc}
 & (1,1) & (2,2) \\
P = \quad (1,1) & p & 1-p \\
(2,2) & 1 & 0
\end{array}
$$

For this network, $D=D^*$ and $G=G^*$. Consider the passage time which starts when a job enters the center 1 queue upon completion of service at center 2 and terminates the next such time at which the job joins the center 1 queue. Also consider the passage time which starts when a job joins the center 1 queue upon completion of service at center 2 and terminates when the job next joins the center 2 queue. For these passage times, the subsets A_1 and A_2 of G are

$$A_1 = \{(i,N): 0 \le i < N\}$$

and

$$A_2 = \{(i,1): 0 < i \le N\}.$$

Then the set S corresponding to the starts of passage times for the marked job is

$$S = \{(i,N,i + 1,1): 0 \le i < N\}.$$

The subset $S' = \{(0,N,1,1)\}$ and the set $A_2' = \{(1,1)\}$.

(4.5) EXAMPLE. Consider a network with two service centers and two job classes such that the set C of (center, class) pairs is C = $\{(1,1),(2,1),(2,2)\}$ and jobs of class 2 have preemptive priority over jobs of class 1 at center 2. Let $Z(t) = (Q_1(t),C_1^{(2)}(t),C_2^{(2)}(t), S_2(t))$. Also suppose that the irreducible routing matrix P is

$$
\begin{array}{cccc}
 & (1,1) & (2,1) & (2,2) \\
(1,1) & 0 & 1 & 0 \\
P = \quad (2,1) & 0 & 0 & 1 \\
(2,2) & 1 & 0 & 0
\end{array}
$$

and observe that $D^* -D \neq \phi$. Assume that for the passage time of interest, the subset A_1 and A_2 of G are

$$
A_1 = \{(q,c_1,c_2,s,N) \in G: s = 2\}
$$

and

$$
A_2 = \{(q,c_1,c_2,s,1) \in G: q > 0\}.
$$

Thus, a passage time starts when a job of class 2 completes service at center 2 and joins the center 1 queue (as class 1). For this network $z_2^* = (0,N-1,0,1)$ and $U(z_2^*) = \{(2,1)\}$. By Proposition (4.3), state $z = (0,N-1,0,2) \in D$ since $q(z; z_2^*,u) = 1$ with $u=(2,1)$. It follows that $\{(0,N-1,0,2,N,1,N-2,0,2,1)\} \in S'$ and can serve as a return state for the passage time simulation.

5. SIMULATION RESULTS

In this section we report simulation results for the data base management system model of Example (2.9). We display point estimates and confidence intervals for the fraction of the time that each of the service centers are busy along with the expected value and percentiles of a limiting passage time denoted by R. This passage time is specified by four subsets A_1, A_2, B_1 and B_2 of G given by

$$
A_1 = \{N- (i+1),0,0,0,0,0,7,i,N - i): 0 \leq i < N\}
$$

and

$$
A_2 = \{N- (i+1),0,0,0,0,0,6,i,N - i): 0 \leq i < N\}
$$

with $B_1 = A_1$ and $B_2 = A_2$. The passage time starts when a job completes service at center 1 as class 7 and terminates when the job next completes service at center 1 as class 7. For this passage time the subset S' of Equation (2.11) is

$$S' = \{(z_0, N, z_0', N): z_0 = (N - 1, 0, 0, 0, 0, 0, 7, 0), \; z_0' = (N-1, 0, 0, 0, 0, 0, 6, 0)\}.$$

Thus there is one single state, (z_0, N), for the passage time. In state (z_0, N) all N jobs are of class 7 at center 1 and the marked job is in service.

The estimates in Tables 1 and 2 for passage time characteristics were obtained using the marked job method of Section 4 and the linear congruential uniform random number generator described by Lewis, Goodman and Miller [15]. Exponential service times were generated by logarithmic transformation of the uniform random numbers; independent streams of exponential random numbers (resulting from different seeds of the uniform random number generator) were used to generate exponential service time sequences for the individual job classes.

In Tables 1 and 2 all service times are exponentially distributed. The mean, λ_j^{-1}, of the service time distribution depends on the class, j, of the job in service. Under these assumtpions, the theoretical value for $E\{R\}$ can be obtained. This value is given in parentheses in Table 1. (Theoretical values for percentiles of R cannot be obtained by these methods). The initial state for the augmented job stack process X (and return state identifying cycles) is the state x' = (N-1, 0, 0, 0, 0, 0, 6, 0, N). For N=2 jobs, a total of 10436 state transitions of the augmented job stack process were required for 200 cycles. A total 427 passage times for the marked job were observed during these 200 cycles. The resulting point estimate for $E\{R\}$ was 78.8978 and the half-length of the 90% confidence interval was 8.0073. For 200, 400, 600,800, and 1000 cucles, the confidence intervals contain the theoretical value. Estimates for N=4 jobs are given in Table 2. Comparison with Table 1 gives an indication of the effect on computational and statistical efficiency of the increase in the number of jobs. For simulations of equal length, the accuracy of the estimates for $E\{R\}$ is roughly comparable.

Table 3 gives point estimates and 90% confidence intervals for the expected value and percentiles of the passage time R when service times to jobs of class 7 are exponentially distributed and service times to the other job classes are constant. Parameter values are as in Table 1. Although the positivity hypothesis used in the proof of Proposition (3.1) is not satisfied, it can be shown that the process $\{X(S_n):n\geq 0\}$ for this network hits x' infinitely often with probability one and that the marked job method of Section 4 is valid.

6. THE LABELLED JOBS METHOD

The labelled jobs method provides estimates for passage times which correspond to passage through a subnetwork of a given network of queues. With the labelled jobs method, observed passage times for all the jobs are used to construct point and interval estimates. Label the jobs from 1 to N and for $t\geq 0$ denote by $N^n(t)$ the label of the job in position n of the job stack at time t, $1\leq n\leq N$. Set

$$X^0(t) = (Z(t),N^1(t),\ldots,N^N(t))$$

and call $X^0 = \{X^0(t):t\geq 0\}$ the *fully augmented job stack process*. The sequence $\{P_n^0:n\geq 1\}$ of passage times for all the jobs enumerated in termination order converges in distribution to a random variable P^0. Moreover, $P^0 = P$, the limiting passage time for any marked job. The goal of the simulation is the estimation of

$$r(f) = E\{f(P)\},$$

where f is a real-values (measurable) function. We assume that $E\{|f(P)|\}<\infty$ and $P\{P \in D(f)\} = 0$, where D(f) is the set of discontinuities of the function f.

Analogous to the set

$$S = \{(k,m):k \in A_1, m \in A_2 \text{ and } p(m;k,u)>0 \text{ for some } u \in U(k)\}$$

of Equation (2.10), define a set T according to

$$T = \{(k,m): k \in B_1, m \in B_2 \text{ and } p(m;k,u) > 0 \text{ for some } u \in U(k)\} \qquad (6.1)$$

For $(k,m) \in T$ the entrances of the augmented job stack process X to state m from state k correspond to the terminations of passage times for the marked job. The labelled jobs method applies to passage times through a subnetwork, i.e., to passage times for which $S \cap T = \phi$

An element z of the set D is called a *single state of the job stack process* for the passage time specified by the sets A_1, A_2, B_1 and B_2 if (i) there exists a state z_1 such that a passage time for some job terminates when the job stack process jumps from z_1 to z and (ii) when the job stack process is in state z no passage times are underway and all jobs are at the same center with exactly one job in service; see [20] for a formal definition. We assume that a single state of the job stack process exists.

Select a single state, z_0, of the job stack process and an initial state $(z_0, n^1, \ldots, n^N)$ for the fully augmented job stack process X^0. Let T_n^0 be the termination time of P_n^0, $n \geq 1$. Denote by $\{\beta_k^0 : k \geq 1\}$ the indices of the successive passage times (irrespective of job identity) which terminate with the job stack process in state z_0. Let $T_0^0 = \beta_0^0 = 0$. Carry out the simulation of the process X^0 in blocks defined by the successive epochs $\{T_{\beta_k^0}^0 : k \geq 1\}$ at which a passage time terminates with the job stack process in state z_0. Set

$$Y_m^0(f) = \sum_{j = \beta_{m-1}^0 + 1}^{\beta_m^0} f(P_j^0)$$

and $\alpha_m^0 = \beta_m^0 - \beta_{m-1}^0$, $m \geq 1$.

It can be shown ([20], Propositions (3.2) and (3.4)) that $P\{Z(T_n^0) = z$ infinitely often$\} = 1$ for any single state z of the job stack process and that the process $\{(Z(T_n^0), P_{n+1}^0) : n \geq 0\}$ is a regenerative process in discrete time.

Moreover, the expected time between regeneration points is finite. It follows that the pairs of random variables $\{(Y_m^0(f), \alpha_m^0) : m \geq 1\}$ are independent and identically distributed, and since $E\{|f(P^0)|\} < \infty$ by assumption, that

$$E\{f(P^0)\} = \frac{E\{Y_1^0(f)\}}{E\{\alpha_1^0\}}$$

Then from a fixed number, n, of blocks of standard regenerative method provides the strongly consistent point estimate

$$\hat{r}_n^0(f) = \frac{\bar{Y}_n^0(f)}{\bar{\alpha}_n^0} = \sum_{m=1}^{n} Y_m^0(f) \bigg/ \sum_{m=1}^{n} \alpha_m^0$$

for $r(f)$. Confidence intervals for $r(f)$ are based on the c.l.t.

$$\frac{n^{1/2} \{\hat{r}_n^0(f) - r(f)\}}{\sigma^0(f)/E\{\alpha_1^0\}} \Longrightarrow N(0,1) \tag{6.2}$$

as $n \to \infty$, where $(\sigma^0(f))^2$ (assumed finite) is the variance of $Y_1^0(f) - r(f)\alpha_1^0$.

7. CONCLUDING REMARKS

The requirements for applicability of the marked job method of Section 4 are that (i) there exist sets D and G as in Equations (2.5) and (2.6), (ii) the sets A_1, A_2, B_1, and B_2 which define the passage time are subsets of G, and (iii) there is recurrence in the sense of Proposition (3.1). We have shown that when all service time density functions are continuous and positive on $(0, \infty)$, a sufficient condition for applicability of the marked job method is that some service center sees only one job class or the lowest priority job class seen by a center be subject to preemption. The requirement that there exist a "single state" (as in Equation (2.11)) is essential.

We have assumed that any preemption of service at a center is of the preemptive-repeat (rather than preemptive-resume) type. This avoids "zero speeds" in the generalized semi-Markov process used to establish recurrence of the augmented job stack process (cf. [20], p. 632). We conjecture that Proposition (3.1) holds for networks with states in which some clocks run at zero speed, provided that the service time density functions satisfy the positivity hypothesis. This would make it possible to handle preemptive-resume networks.

The marked job method prescribes observation of passage times for an arbitrarily chosen, distinguished job. With the broadly applicable marked job method the half-length of the confidence interval (obtained from a simulation of fixed length) for the expected value of a general function f of the limiting passage time is proportional to a certain quantity e(f).The labelled jobs method provides estimates for passage times through a subnet-work. With the labelled jobs method, observed passage times for all the jobs are used to construct point and interval estimates and (with the same constant of proportionality) the half-length of the confidence interval is proportional to a quantity $e^0(f)$. Since these quantities are independent of the blocks of the underlying regenerative process, they are appropriate measures of the statistical efficiency of the estimation procedures. For Markovian networks of queues, it is possible to compute theoretical values for expected passage times and the associated variance constants appearing in central limit theorems used to form confidence intervals; here f is the identity function. This leads to a quantitative assessment ([9]) of the relative statistical efficiencies of the estimation procedures (for expected passage times) in [8] for networks with Cox-phase service times.

For networks of queues with general service times, there is little hope of computing the needed theoretical values, even for expected passage times. Using central limit theorem and continuous mapping theorem arguments, it can be shown ([10], Proposition (4.12)) that for any function f (and all numbers of jobs in the network) $e^0(f) \le e(f)$; i.e., the confidence intervals constructed using the labelled jobs methods are shorter than those obtained from the marked job method. This is consistent with intuition since the labelled jobs method extracts more passage time information from a fixed length simulation run.

TABLE 1

Simulation Results for Passage Time R in DL/I Component Model

$N=2, \psi_1=0.1, \psi_2=0.2, \lambda_1^{-1}=50, \lambda_2^{-1}=\lambda_4^{-1}=3.3, \lambda_3^{-1}=\lambda_5^{-1}=1.5, \lambda_6^{-1}=6.7, \lambda_6^{-1}=1.0$

Exponential Service Times for All Job Classes

Return State is $(1,0,0,0,0,6,0,2)$

No. of cycles, n	200	400	600	800	1000
Total simulated time	33689.38	80632.23	129362.60	174804.60	219304.30
No. of transitions/cycle	52.18	59.98	63.35	64.33	64.47
Fraction of time center 1 busy (0.7498)	0.7770	0.7528	0.7433	0.7480	0.7484
	0.0524	0.0289	0.0219	0.0180	0.0157
Fraction of time center 2 busy (0.5913)	0.5282	0.5747	0.5894	0.5837	0.5839
	0.0603	0.0321	0.0241	0.0199	0.0174
$\bar{\alpha}_n$	2.135	2.400	2.513	2.541	2.589
E{R} (84.556)	78.8978	83.9919	85.7842	85.9836	84.7062
	8.0073	5.3648	4.3583	3.5606	3.0340
$P\{R\leq 21.14\}$	0.1897	0.2010	0.2042	0.1899	0.1943
	0.0324	0.0226	0.0189	0.0157	0.0139
$P\{R\leq 42.28\}$	0.4145	0.4177	0.4178	0.4097	0.4125
	0.0416	0.0281	0.0217	0.0188	0.0165
$P\{R\leq 84.56\}$	0.7424	0.6990	0.6916	0.6837	0.6840
	0.0337	0.0246	0.0198	0.0170	0.0149
$P\{R\leq 169.11\}$	0.8829	0.8615	0.8561	0.8608	0.8617
	0.0258	0.0182	0.0148	0.0124	0.0111
$P\{R\leq 338.22\}$	0.9742	0.9729	0.9675	0.9685	0.9718
	0.0142	0.0092	0.0081	0.0067	0.0056

TABLE 2

Simulation Results for Passage Time R in DL/I Component Model

$N=4, \psi_1=0.1, \psi_2=0.2, \lambda_1^{-1}=50, \lambda_2^{-1}=\lambda_4^{-1}=3.3, \lambda_3^{-1}=\lambda_5^{-1}=1.5, \lambda_6^{-1}=6.7, \lambda_7^{-1}=1.0$

Exponential Service Times for All Job Classes

Return State is (3,0,0,0,6,0,4)

No. of cycles, n	200	400	600	800	1000
Total simulated time	145268.80	270127.30	406800.50	537384.00	683064.00
No. of transitions/cycle	249.06	231.85	233.82	232.62	236.36
Fraction of time center 1 busy	0.8755 / 0.0185	0.8758 / 0.0141	0.8801 / 0.0111	0.8849 / 0.0092	0.8817 / 0.0084
Fraction of time center 2 busy	0.7263 / 0.0266	0.7141 / 0.0197	0.7129 / 0.0156	0.7122 / 0.0132	0.7183 / 0.0116
$\bar{\alpha}_n$	4.930	4.658	4.745	4.702	4.754
E{R}	147.3315 / 7.3648	144.9959 / 5.6797	142.8874 / 4.3073	142.8453 / 3.6739	143.6820 / 3.2584
$P\{R \leq 21.14\}$	0.0923 / 0.0201	0.0886 / 0.0139	0.0861 / 0.0109	0.0829 / 0.0095	0.0852 / 0.0082
$P\{R \leq 42.28\}$	0.2039 / 0.0271	0.1997 / 0.0201	0.1939 / 0.0157	0.1877 / 0.0140	0.1906 / 0.0119
$P\{R \leq 84.56\}$	0.4320 / 0.0272	0.4283 / 0.0224	0.4254 / 0.0174	0.4242 / 0.0152	0.4222 / 0.0131
$P\{R \leq 169.11\}$	0.6907 / 0.0252	0.7112 / 0.0180	0.7201 / 0.0143	0.7188 / 0.0124	0.7137 / 0.0112
$P\{R \leq 338.22\}$	0.9077 / 0.0164	0.9152 / 0.0111	0.9203 / 0.0085	0.9203 / 0.0072	0.9178 / 0.0066

TABLE 3

Simulation Results for Passage Time R in DL/I Component Model

$N=2, \psi_1=0.1, \psi_2=0.2, \lambda_1^{-1}=50, \lambda_2^{-1}=\lambda_4^{-1}=3.3, \lambda_3^{-1}=\lambda_5^{-1}=1.5, \lambda_6^{-1}=6.7, \lambda_7^{-1}=1.0$

Exponential Service Times for Class 7 Jobs

Constant Service Times for Other Job Classes

Return State is $(1,0,0,0,0,6,0,2)$

No. of cycles, n	200	400	600	800	1000
Total simulated time	45291.23	92218.60	143658.30	193607.10	241766.60
No. of transitions/cycle	68.81	67.39	69.75	70.28	70.22
Fraction of time center 1 busy	0.7727 0.0256	0.7425 0.0205	0.7386 0.0171	0.7360 0.0149	0.7360 0.0135
Fraction of time center 2 busy	0.6071 0.0312	0.6193 0.0229	0.6190 0.0196	0.6198 0.0169	0.6195 0.0152
$\bar{\alpha}_n$	2.795	2.762	2.765	2.751	2.755
$E\{R\}$	81.0219 6.4320	83.4557 4.4219	83.5933 3.6403	87.9632 3.2392	87.7556 2.8777
$P\{R<21.14\}$	0.1843 0.0299	0.1828 0.0206	0.1706 0.0170	0.1699 0.0147	0.1699 0.0130
$P\{R<42.28\}$	0.4741 0.0370	0.4588 0.0250	0.4394 0.0203	0.4321 0.0181	0.4276 0.0161
$P\{R<84.56\}$	0.7048 0.0303	0.6914 0.0217	0.6745 0.0185	0.6674 0.0162	0.6708 0.0144
$P\{R<169.11\}$	0.8712 0.0228	0.8606 0.0157	0.8517 0.0134	0.8487 0.0117	0.8505 0.0103
$P\{R<338.22\}$	0.9696 0.0122	0.9692 0.0086	0.9668 0.0070	0.9668 0.0060	0.9673 0.0053

ACKNOWLEDGEMENT : The authors are both grateful to the National Science Foundation for support under Grant MCS-8203483. In addition, Donal L. Iglehart gratefully acknowledges partial support under Office of Naval Research Contract N00014-76-C-0578 (NR 042-343).

REFERENCES

[1] Cox, D.R. (1955). A use of complex probabilities in the theory of queues. Proc. Cambridge Philos. Soc. 51, 313-319.

[2] Crane, M.A. and Iglehart, D.L. (1975). Simulating stable stochastic systems: III, Regenerative processes and discrete event simulation. Operations Res. 23, 33-45.

[3] Crane, M.A. and Lemoine, A.J. (1977). An Introduction to the Regenerative Method for Simulation Analysis. Lecture Notes in Control and Information Sciences, Vol. 4, Springer-Verlag, Berlin, Heidelberg, New York.

[4] Fishman, G.S. (1978). Principles of Discrete Event Simulation. John Wiley, New York.

[5] Fossett, L.D. (1979). Simulating generalized Semi-Markov process. Technical Report No. 2. Department of Operations Res., Stanford, Wisconsin.

[6] Glynn, P.W. (1983). Forthcoming technical report. Department of Industrial Engineering. University of Wisconsin, Madison, Wisconsin.

[7] Heidelberger, P. and Welch, P.D. (1981). A spectral method for confidence, interval generation and run length control in simulations. Comm. Assoc. Comput. Mach. 24, 233-245.

[8] Iglehart, D.L. and Shedler, G.S. (1980). Regenerative Simulation of Response Times in Networks of Queues. Lecture Notes in Control and Information Sciences, Vol. 26. Springer-Verlag, Berlin, Heidelberg, New York.

[9] Iglehart, D.L. and Shedler, G.S. (1981). Regenerative simulation of response times in networks of queues: statistical efficiency. Acta Informatica 15, 347-363.

[10] Iglehart, D.L. and Shedler, G.S. (1983). Statistical efficiency of regenerative simulation methods for networks of queues. Adv. Appl. Probability 15, 183-197.

[11] König, D., Matthes, K. and Nawrotzki, K. (1967). Verallgemeinerungen der Erlangschen und Engsetschen Formeln. Akademie-Verlag, Berlin.

[12] König, D., Matthes, K. and Nawrotzki, K. (1974). Unempfindlichkeitseigenschaften von Bedienungsprozessen. Appendix to Gnedenko, B.V. and Kovalenko, I.N., Introduction to Queueing Theory, German edition.

[13] Lavenberg, S.S. and Shedler, G.S.(1976). Stochastic modelling of processor scheduling with application to data base amangement systems. IBM J. Res. Develop. 19, 437-448.

[14] Law, A.M. and Carson, J. (1979). A sequential procedure for determining the length of a steady-state simulation. Operations Res. 27, 1011-1025.

[15] Lewis, P.A.W., Goodman, A.S. and Miller, J.M. (1969). A pseudo-random number generator for the System/360. IBM Systems J. 8, 199-220.

[16] Matthes, K. (1962). Zur Theorie der Bedienungsprozesse. Trans. 3rd Prague Conference on Information Theory and Statistical Decision Functions. Prague.

[17] Miller, D.R. (1972). Existence of limits in regenerative processes. Ann. Math. Statist. 43, 1275-1282.

[18] Schmeiser, B. (1982). Batch size effects in the analysis of simulation output. Operations Res. 30, 556-568.

[19] Shedler, G.S. and Slutz, D.R. (1981). Irreducibility in closed multiclass networks of queues with priorities: passage times of a marked job. Performance Eval. 1, 334-343.

[20] Shedler, G.S. and Southard, H. (1982). Regenerative simulation of networks of queues with general service times: passage through subnetworks. IBM J. Res. Develop. 26, 625-633.

[21] Smith, W.L. (1958). Renewal theory and its ramifications. J. Roy. Statist. Soc. Ser. B 20, 243-302.

[22] Whitt, W. (1980). Continuity of generalized semi-Markov processes. Math. Operations Res. 5, 494-501.

SIMULATION USES OF THE EXPONENTIAL DISTRIBUTION

Sheldon M. Ross
Department of Industrial Engineering
and Operations Research
University of California, Berkeley

and

Zvi Schechner
Department of Industrial Engineering
and Operations Research
Columbia University
New York, New York

0. INTRODUCTION AND SUMMARY

In this paper, we show how exponential random variables can be efficiently used in a variety of simulation problems. One of the problems we are concerned with is the simulation of order statistics from a normal population. In Section 1, we discuss the general problem of simulating order statistics and in Section 2, we consider the normal case. We start by showing how the Von-Neumann rejection approach via the exponential can be modified to become an efficient algorithm for generating a normal and then in Section 2.1, we present a method for generating normal order statistics. In Section 3, we show how to use the exponential to efficiently simulate random permutations with weights. In Section 4, we consider the problem of simulating a 2-dimensional Poisson process and in Section 5, we generalize this to allow the process to be nonhomogeneous.

1. SIMULATING ORDER STATISTICS

Consider the problem of simulating the order statistics $X_{(1)} < X_{(2)} < \ldots < X_{(n)}$ from a sample of size n from the continuous distribution F. If F is invertible, this can be accomplished by simulating $U_{(1)} < \ldots < U_{(n)}$ the order statistics from a sample of n uniform $(0,1)$ random variables and then setting $X_{(i)} = F^{-1}(U_{(i)})$. To generate the $U_{(i)}$, we can simulate n uniform $(0,1)$ random variables $U_1, \ldots, U_n$ and then order them. (It should be mentioned at this point that the time necessary for sorting the U_i's is not nearly as large as has been indicated in the simulation literature (see for instance [5]). The reason being that the sorting need not be done via a general purpose procedure (such as quicksort which requires on the order of $n \log n$ comparisons) but rather can be done with a procedure that utilizes the fact that the values to be sorted are generated from a uniform distribution (see 5.2.1 of [2]). This leads to an algorithm requiring on the order of n comparisons.) We now present another approach for generating $U_{(1)}, \ldots, U_{(n)}$.

Let $X_1, \ldots, X_{n+1}$ be independent exponential random variables with rate 1 and interpret X_i as the ith interarrival time of a Poisson process. Now set $S_i = \sum_{j=1}^{i} X_j$ and so S_i is the time of the ith event. Now apply the well known result that given S_{n+1}, $S_1, \ldots, S_n$ are distributed as the ordered values of a set of n uniform $(0, S_{n+1})$ random variables. Hence, $\frac{S_1}{S_{n+1}}, \ldots, \frac{S_n}{S_{n+1}}$ have the joint distribution of $U_{(1)}, \ldots, U_{(n)}$. That is, we have the following algorithm.

Step 1: Generate $n + 1$ independent exponential random variables $X_1, \ldots, X_{n+1}$.

Step 2: Set $U_{(i)} = \sum_{j=1}^{i} X_j \Big/ \sum_{j=1}^{n+1} X_j$, $i = 1, \ldots, n$.

Remarks:

The above procedure was discussed in [4]. The exponential random variables can be generated either by using $X_i = -\log U_i$, U_i being uniform on $(0,1)$, or by using any of the other known algorithms.

Suppose now, as in [6], that of the n order statistics, $U_{(1)}, \ldots, U_{(n)}$ we only desire $U_{(i)}, U_{(i+1)}, \ldots, U_{(i+k)}$. To simulate this set of $k + 1$ of the n order statistics, start by simulating X, a gamma random variable with parameters $(i,1)$ (either by summing exponentials when i is small or by using one of the algorithms that is more efficient than this when i is large). Interpret X as the time of the ith event of a Poisson process with rate 1. Now, simulate k exponential random variables with rate 1--call them $X_1, \ldots, X_k$--and set $S_{i+j} = X + \sum_{\ell=1}^{j} X_\ell$, $j = 1, \ldots, k$. Now simulate Y, a gamma random variable with parameters $(n + 1 - i - k, 1)$, which will represent the time between the $(i + k)$th and the $(n + 1)$st event. Then

$$\frac{X}{S_{i+k} + Y}, \frac{S_{i+1}}{S_{i+k} + Y}, \ldots, \frac{S_{i+k}}{S_{i+k} + Y}$$ will have the desired distribution.

That is, we have the following algorithm for simulating $U_{(i)}, \ldots, U_{(i+k)}$.

Step 1: Generate $k + 2$ independent random variables $X, Y, X_1, \ldots, X_k$ with X being gamma with parameters $(i,1)$, Y being gamma with parameters $(n + 1 - i - k, 1)$, and X_i being exponential with rate 1.

Step 2: Set for $j = 0,1, \ldots, k$

$$U_{(i+j)} = \frac{X + \sum_{\ell=1}^{j} X_\ell}{X + \sum_{\ell=1}^{k} X_\ell + Y}$$

Even easier than simulating uniform order statistics is simulating exponentially distributed ones. Let $Y_1, \ldots, Y_n$ be independent exponential random variables with rate 1 and set

$$Y_{(1)} = \frac{Y_1}{n}$$

$$Y_{(2)} = Y_{(1)} + \frac{Y_2}{n - 1}$$

$$Y_{(j)} = Y_{(j-1)} + \frac{Y_j}{n - j + 1}$$

$$Y_{(n)} = Y_{(n-1)} + Y_n$$

It then follows from the lack of memory property of exponentials plus the fact that the minimum of independent exponentials is exponentially distributed with rate equal to the sum of the rates of the exponentials over which we are minimizing, that $Y_{(1)}, \ldots, Y_{(n)}$ are distributed as the order statistics of a set of n independent exponentials with rate 1.

Since $F^{-1}(1 - e^{-Y})$ and $F^{-1}(e^{-Y})$ both have distribution F when Y is exponential with rate 1, it follows that we can simulate a set of n order statistics from F by first simulating the exponential order statistics $Y_{(1)} \ldots Y_{(n)}$ and then either setting $X_{(i)} = F^{-1}(1 - e^{-Y}{(i)})$, $i = 1, \ldots, n$ or setting $X_{(i)} = F^{-1}(e^{-Y}{(n+1-i)})$, $i = 1, \ldots, n$.

Hence, if $F^{-1}(1 - e^{-x})$ or $F^{-1}(e^{-x})$ is as easy or almost as easy to calculate as $F^{-1}(x)$ it is more efficient to first simulate exponential rather than uniform order statistics.

2. SIMULATING NORMAL RANDOM VARIABLES

Let Z denote a unit normal random variable and set $X = |Z|$. The Von-Neumann Rejection method for simulating X , via the exponential distribution, leads to the following well-known algorithm.

(a) Generate independent exponentials with rate 1, Y_1 and Y_2 .

(b) Set $X = Y_1$ if $Y_2 \geq (Y_1 - 1)^2/2$. Otherwise return to (a).

To improve on this note that, by the lack of memory property of the exponential, it follows that if we reject in Step (b)--that is, if $Y_2 < (Y_1 - 1)^2/2$ or, equivalently, if $Y_1 > 1 + \sqrt{2Y_2}$--then $Y_1 - 1 - \sqrt{2Y_2}$ is exponential with rate 1. Hence, we can use this as one of the exponential values in the next iteration. In fact, even if we accept in (b), we can generate an exponential (independent of X) by computing $Y_2 - (Y_1 - 1)^2/2$ which will be exponential with rate 1.

Hence, summing up, we have the following algorithm which generates an exponential with rate 1 and an independent unit normal random variable.

Step 1: Generate Y_1 , an exponential random variable with rate 1.

Step 2: Generate Y_2 , an exponential with rate 1.

Step 3: If $Y_2 - (Y_1 - 1)^2/2 > 0$ set $Y = Y_2 - (Y_1 - 1)^2/2$ and go to Step 4. Otherwise reset Y_1 to equal $Y_1 - 1 - \sqrt{2Y_2}$ and return to Step 2.

Step 4: Generate a random number U and set

$$Z = \begin{cases} Y_1 & \text{if } U \leq 1/2 \\ -Y_1 & \text{if } U > 1/2 . \end{cases}$$

The random variables Z and Y generated by the above are independent with Z being normal with mean 0 and variance 1 and Y being exponential with rate 1. (If we want the normal random variable to have mean μ and variance σ^2 , just take $\mu + \sigma Z$).

Remarks:

(1) As is well known, the above requires a geometric distributed number of iterations of Step 2 with mean $\sqrt{2e/\pi} \approx 1.32$.

(2) The final random number of Step 4 need not be separately simulated but rather can be obtained from the first digit of any random number used earlier. That is, suppose we generate a random number to simulate an exponential, then we can strip off the initial digit of this random number and just use the remaining digits (with the decimal point moved one step to the right) as the random number. If this initial digit is 0,1,2,3 or 4 (or 0 if the computer is generating binary

digits), then we take the sign of Z to be positive and take it negative otherwise.

(3) If we are generating a sequence of unit normal random variables, then we can use the exponential obtained in Step 4 as the initial exponential needed in Step 1 for the next normal to be generated. Hence, on the average, we can simulate a unit normal by generating 1.32 exponentials and computing 1.32 squares and .32 square roots.

2.1 Simulating Normal Order Statistics

Suppose we want to simulate the order statistics of n independent unit normal random variables. To do so we will generate m independent exponentials with rate 1--$W_1, W_2, \ldots, W_m$--and use the above algorithm to generate a binomial distributed number of normals. However, so as to eliminate the need to order these resultant normals we shall first order the exponentials. We thus have the following approach:

Step 1: Generate independent exponential random variables with rate 1, $W_1, \ldots, W_m$ and set

$$W_{(1)} = W_1/m$$

$$W_{(j)} = W_{(j-1)} + \frac{W_j}{m - j + 1} \quad , \quad j = 2, \ldots, m \ .$$

Step 2: Set $k = 0$, $\ell = 1$.

Step 3: Generate Y an exponential random variable with rate 1.

Step 4: If $k = m$ stop; otherwise reset k to equal $k + 1$.

Step 5: If $Y - \dfrac{(W_{(k)} - 1)^2}{2} \geq 0$ set $X_{(\ell)} = W_{(k)}$, reset Y to equal $Y - \dfrac{(W_{(k)} - 1)^2}{2}$,

reset ℓ to equal $\ell + 1$ and go to Step 4. Otherwise if $Y - \dfrac{(W_{(k)} - 1)^2}{2} < 0$ go to Step 3.

The result of the above will be the random variables $X_{(1)}, \ldots, X_{(R)}$ where $R = \ell - 1$ is binomially distributed with parameters m and $p = \sqrt{\pi/2e}$. Since the above is equivalent to using the rejection method on the m independent and identically distributed (iid) random variables $W_1, \ldots, W_m$ it follows that the accepted variables are also iid. Hence, given R , $X_{(1)}, \ldots, X_{(R)}$ are distributed as the order statistics of a set of R random variables having the distribution of $|Z|$ where Z is a unit normal.

However, as R need not equal n we are not finished. If $R > n$, then we need randomly choose $R - n$ of the indices $1, \ldots, R$ and then eliminate the variables having these indices. For instance if $n = 5$ and $R = 7$ and the random choice of 2 of the first 7 integers is 1 and 4 then the resultant 5 order statistics are $\bar{\bar{X}}_{(i)}$, $i = 1, \ldots, 5$ where $\bar{\bar{X}}_{(1)} = X_{(2)}$, $\bar{\bar{X}}_{(2)} = X_{(3)}$, $\bar{\bar{X}}_{(3)} = X_{(5)}$, $\bar{\bar{X}}_{(4)} = X_{(6)}$, $\bar{\bar{X}}_{(5)} = X_{(7)}$. The best way to randomly choose $R - n$ of the indices $1, \ldots, R$ depends on the relative size of $R - n$ in comparison to R. If $R - n$ is small compared to R, then we can generate random numbers U and eliminate the index $[RU] + 1$ continuing until $R - n$ distinct integers have been removed. If $R - n$ is moderate in comparison to R, then there is also a very simple algorithm for the selection (see Section 3.4.2 of [1]).

If $R < n$, then we can reapply the above steps (with a different value for m) and then merge the two sets of order statistics. If this results in greater than n order statistics, then we use the above technique to randomly eliminate certain of them. If the combined set is still of size less than n, we again reapply the above steps (with again a different value for m) until finally we have at least n order statistics.

The problem of choosing the appropriate value of m when k order statistics need be generated remains. As $E[R] = .7576\, m$ and $\sqrt{\text{Var } R} = .4285 \sqrt{m}$, we can be almost certain of having $R \geq k$ if we take m to be approximately equal to $\frac{4}{3} k + \sqrt{4k/3}$. However, perhaps a value around $\frac{4}{3} k$ (which would give one roughly a fifty-fifty chance of having $R \geq n$) would work out better. Numerical experience will be necessary to determine the best m when the order statistics of k of the variables are needed.

Once the above is accomplished, we will have the order statistics of the absolute values of n unit normals. To obtain the order statistics of n unit normals generate n uniforms—one for each order statistics—and break the list of order statistics of the absolute normals into 2 parts—those having a U-value less than 1/2 and those having it greater than 1/2—keeping the same relative ordering within the 2 lists. The reversed order of the second list, with each element in this list given a negative sign, followed by the first list now yields the order statistics of n unit normals.

Remark:

As in Section 2, the final n uniforms needed to determine the signs of the unit normals need not actually be generated but can be "stripped off" from random numbers used earlier.

3. SIMULATING RANDOM PERMUTATIONS WITH WEIGHTS

Consider an urn containing n balls--the ith of which weighs w_i, $i = 1, \ldots,$ n . The balls are sequentially removed, without replacement, from the urn according to the following scheme. At each withdrawal, a given remaining ball will be removed with probability equal to its weight devided by the sum of the weights of the remaining balls. We are interested in simulating the random permutation $\underline{I} = (I_1, \ldots, I_n)$ where I_i is the index of the ith ball to be removed.

The following algorithm for simulating I is suggested,

Step 1: Simulate independent exponential random variables $E_1, \ldots, E_n$ each having rate 1.

Step 2: Set $X_i = E_i/w_i$, $i = 1, \ldots,$ n .

Step 3: Let $I = (I_1, \ldots, I_n)$ where I_j is the index of the jth smallest of $X_1, \ldots, X_n$.

That the above algorithm indeed works follows from the lack of memory property of the exponential distribution in conjunction with the fact that if X_i, $i \geq 1$, are independent exponentials with respective rates w_i, then

$$P\left\{X_{i_1} = \min\left(X_{i_1}, X_{i_2}, \ldots, X_{i_k}\right)\right\} = \frac{w_{i_1}}{\sum_{j=1}^{k} w_{i_j}} \; .$$

Remark:

The above can also be used in the problem of [5] which is concerned with generating a weighted random sample of size k of the n balls.

4. SIMULATING A CIRCULAR REGION OF A TWO-DIMENSIONAL POISSON PROCESS

A point process consisting of randomly occurring points in the plane is said to be a two-dimensional Poisson process having rate λ if

(a) the number of points in any given region of area A is Poisson distributed with mean λA ; and

(b) the number of points in disjoint regions are independent.

For a given fixed point $\underline{0}$ in the plane, we now show how to simulate events occurring according to a two dimensional Poisson process with rate λ in a circular region of radius r centered about $\underline{0}$. Let R_i , $i \geq 1$, denote the distance between $\underline{0}$ and its ith nearest Poisson point, and let $C(a)$ denote the circle of radius a centered at $\underline{0}$. Then

$$P\left\{\pi R_1^2 > b\right\} = P\{\text{no points in } C(\sqrt{b/\pi})\} = e^{-\lambda b} \ .$$

Also,

$$P\left\{\pi R_2^2 - \pi R_1^2 > b \mid R_1 = r\right\}$$

$$= P\left\{R_2 > \sqrt{(b + \pi r^2)/\pi} \mid R_1 = r\right\}$$

$$= P\left\{\text{no points in } C\left(\sqrt{(b + \pi r^2)/\pi}\right) - C(r)\right\} = e^{-\lambda b} \ .$$

In fact, the same argument can be repeated to obtain

Proposition 1: With $R_0 = 0$,

$\pi R_i^2 - \pi R_{i-1}^2$, $i \geq 1$, are independent exponentials with rate λ .

In other words, the amount of area that need be traversed to encompass a Poisson point is exponential with rate λ . Since, by symmetry, the respective angles of the Poisson points are independent and uniformly distributed over $(0, 2\pi)$ we thus have the following algorithm for simulating the Poisson process over a circular region of radius r about $\underline{0}$.

Step 1: Generate independent exponentials with rate 1, $X_1, X_2, \ldots,$ stopping at

$$N = \min\left\{n : \frac{X_1 + \ldots + X_n}{\lambda \pi} > r^2\right\} \ .$$

Step 2: If $N = 1$, stop, there are no points in $C(r)$. Otherwise, for $i = 1, \ldots,$ $N - 1$ set

$$R_i = \sqrt{(X_1 + \ldots + X_i)/\lambda\pi}$$

Step 3: Generate independent uniform $(0,1)$ random variables $U_1, \ldots, U_{N-1}$.

Step 4: Return the $N - 1$ Poisson points in $C(r)$ whose polar coordinates are

$$(R_i, 2\pi U_i) \quad , \quad i = 1, \ldots, N - 1 .$$

The above algorithm which requires, on average, $1 + \lambda\pi r^2$ exponentials and an equal number of uniform random numbers can be compared with the procedure suggested in [3]. This latter procedure simulates points in $C(r)$ by first simulating N , the number of such points and then uses the fact that, given N , the points are uniformly distributed in $C(r)$. Hence, the procedure of [3] requires the simulation of N , a Poisson random variable with mean $\lambda\pi r^2$ and must then simulate N uniform points on $C(r)$, by simulating R from the distribution $F_R(a) = a^2/r^2$ and θ from uniform $(0,2\pi)$ and must then sort these N uniform values in increasing order of R . The main advantage of our procedure is that it eliminates the need to sort.

The above algorithm can be thought of as the fanning out of a circle centered at $\underline{0}$ with a radius that expands continuously from 0 to r . The successive radii at which Poisson points are encountered is simulated by noting that the additional area necessary to encompass a Poisson point is always, independent of the past, exponential with rate λ . This technique can be used to simulate the process over noncircular regions. For instance, consider a nonnegative function $g(x)$ and suppose we are interested in simulating the Poisson process in the region between the x-axis and g with x going from 0 to T (see Figure 1).

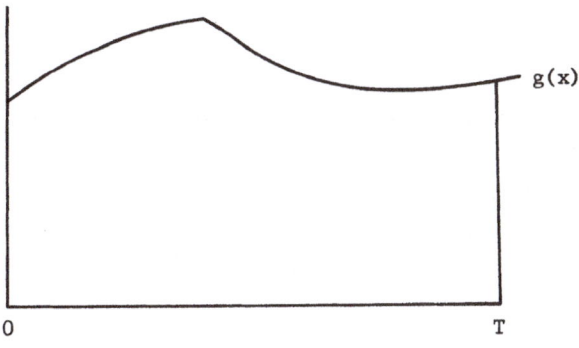

FIGURE 1

To do so we can start at the left hand end and fan vertically to the right by considering the successive areas $\int_0^a g(x)\,dx$. Now if $X_1 < X_2 < \ldots$ denote the successive projections of the Poisson points on the x-axis, then analagous to Proposition 1, it will follow that (with $X_0 = 0$) $\lambda \int_{X_{i-1}}^{X_i} g(x)dx$, $i \geq 1$, will be independent exponentials with rate 1. Hence, we simulate $E_1, E_2, \ldots,$ independent exponentials with rate 1, stopping at

$$N = \min \left\{ n \; : \; E_1 + \ldots + E_n > \lambda \int_0^T g(x)dx \right\}$$

and determine $X_1, \ldots, X_{N-1}$ by

$$\lambda \int_0^{X_1} g(x)dx = E_1$$

$$\lambda \int_{X_1}^{X_2} g(x)dx = E_2$$

$$\vdots$$

$$\lambda \int_{X_{N-2}}^{X_{N-1}} g(x)dx = E_{N-1} \; .$$

If we now simulate $U_1, \ldots, U_{N-1}$--independent uniform $(0,1)$ random numbers, then as the projection on the y-axis of the Poisson point whose x-coordinate is X_i is uniform on $(0, g(X_i))$, it follows that the simulated Poisson points in the interval are $(X_i, U_i g(X_i))$, $i = 1, \ldots, N-1$.

Of course, the above technique is most useful when g is regular enough so that the above equations can be solved for the X_i . For instance, if $g(x) = y$ (and so the region of interest is a rectangle), then

$$X_i = \frac{E_1 + \ldots + E_i}{\lambda y} \; , \quad i = 1, \ldots, N-1$$

and the Poisson points are

$$(X_i, y U_i) \; , \quad i = 1, \ldots, N-1 \; .$$

5. SIMULATING TWO DIMENSIONAL NONHOMOGENEOUS POISSON PROCESSES

Consider now a nonhomogeneous Poisson process with intensity function $\lambda(x,y)$ and suppose we are interested in simulating this process over a region R. Let λ be such that $\lambda(x,y) \leq \lambda$ for all $(x,y) \in R$. A thinning algorithm which first simulates a Poisson process having rate λ over R and then accepts the resulting Poisson point (x,y) with probability $\frac{\lambda(x,y)}{\lambda}$ was recommended in [3]. However, we recommend a conditional approach that first simulates N, a Poisson random variable with mean $|R| = \iint\limits_{(x,y) \in R} \lambda(x,y)\,dxdy$—which we assume is calculable (and which represents the number of points in R)—and then chooses N points in the region R by simulating from the density $f(x,y) = \frac{\lambda(x,y)}{|R|}$. To simulate from this density, an acceptance rejection procedure that simulates a point (X,Y) uniformly in the region and then accepts it with probability $\lambda(X,Y)/\lambda$ can be used. That is, we have the following algorithm.

Step 1: Simulate N, a Poisson random variable with mean $|R|$.

Step 2: Simulate a point (X,Y) uniformly distributed in R and a uniform $(0,1)$ variable U and accept (X,Y) if

$$U \leq \frac{\lambda(X,Y)}{\lambda} \,.$$

If there have been a total of N points that have been accepted, stop—otherwise return to Step 2.

Remarks:

(i) Each random point (X,Y) will be accepted with probability $|R|/\lambda A(R)$ where $A(R) = \iint\limits_{R} dydx$ is the area of R. Hence, the mean number of iterations of Step 2 needed to generate N accepted points is $N\lambda A(R)/|R|$. As N has mean $|R|$, the above thus needs, on average, $\lambda A(R)$ iterations of Step 2.

(ii) The method for simulating (X,Y) uniformly in R depends on the geometry of R. One possibility is to enclose R in a rectangle and thus randomly choose a point (X,Y) in this rectangle and then accept it if $(X,Y) \in R$.

(iii) The fanning out technique of Section 3 can still be employed when $\lambda(x,y)$ is easily integrated and R is "nice." For instance suppose R is the region of Figure 1. Then if $X_1 < X_2 < \ldots$ denote the successive projections of the Poisson points on the x-axis, then with $X_0 = 0$, $\displaystyle\int_{X_{i-1}}^{X_i} \int_0^{g(x)} \lambda(x,y)\,dydx$ are independent exponentials with rate 1. Also given a Poisson point on the line $x = X_i$, the y-coordinate is distributed according to density

$$\lambda(X_i,y) \Bigg/ \int_0^{g(X_i)} \lambda(X_i,y)\,dy \,.$$

REFERENCES

1. Knuth, D., _Semi-Numerical Algorithms_, Volume 2 of _The Art of Computer Programming_, Second Edition, Addison Wesley, 1981.
2. Knuth, D., _Sorting and Searching_, Volune 3 of _The Art of Computer Programming_, Addison Wesley, 1972.
3. Lewis, P. A. W. and G. S. Shedler, "Simulation of Nonhomogeneous Poisson Processes by Thinning," _Naval Res. Log. Quart._, Vol. 26, No. 3, pp. 403-412, 1979.
4. Lurie, D. and H. O. Hurtley, "Machine Generation of Order Statistics for Monte Carlo Simulation," _American Statistician_, Vol. 26, No. 1, pp. 26-27, 1972.
5. Lurie, D. and R. Mason, "Empirical Investigations of Several Techniques for Computer Generation of Order Statistics," _Communications in Statistics_, Vol. 2, pp. 363-371, 1973.
6. Ramberg, J. and P. Tadikamalla, "On the Generation of Subsets of Order Statistics," _J. Statis. Comput. Simul._, Vol. 6, pp. 239-241, 1978.
7. Schucany, W., "Order Statistics in Simulation," _J. Statist. Comput._, Vol. 1, pp. 281-286, 1972.
8. Wong, C. K. and M. C. Easton, _SIAM J. Comput._, Vol. 9, pp. 111-113, 1980.

A PROBABILISTIC ANALYSIS OF MONTE CARLO ALGORITHMS FOR A CLASS OF COUNTING PROBLEMS

A. Frigessi, C. Vercellis
Dipartimento di Matematica, Università di Milano

Abstract

The use of Monte Carlo algorithms for the approximate solution of hard counting problems has been recently proposed by different authors.

In this paper, a general framework for analysing the time complexity of these algorithms is introduced by defining different indeces of their worst case or average behaviour. Then, a Monte Carlo procedure for the approximate solution, within a prefixed accuracy level, of a particular class of #P-hard counting problems (the Intersection Cardinality Problem, denoted as ICP) is considered, and tight bounds on its indeces of performance are derived. It is shown that, under mild assumptions on the parameters of the stochastic model of ICP, the average time of the algorithm is consistently less than its worst case time complexity.

Keywords : Counting Problems, #P-completeness, Monte Carlo Algorithms, Probabilistic Analysis.

1. Introduction

In several combinatorial problems one is required to count the number of elements, belonging to a given ground set, which satisfy a certain property (such elements will be called hereafter *feasible points)*. In fact, *counting problems* have been traditionally considered besides *decision problems,* in which one is asked to decide whether at least one feasible point does exist.

Recently, Valiant $[10]$ introduced the class #P and the associated concept of #P-*completeness* in order to classify counting problems in terms of their computational complexity. A counting problem π belongs to the class #P, if there exists a nondeterministic Turing Machine such that for each instance I of π the number of accepting computations equals the number of feasible points of I and such that, for each I ∈ π, the length of the longest accepting computation is bounded by a polynomial in the size of I. A counting problem π is #P-complete if it belongs to the class #P and if every other problem π' ∈ #P is Turing reducible to π ($[8]$, $[10]$) .

Several relevant counting problems in the area of discrete mathematics and operations research have been shown to be #P-complete ($[1]$, $[8]$, $[9]$, $[10]$, $[11]$). Since it seems unlikely to find polynomial algorithms for their exact solution, the question arises of how to design fast algorithms for their approximate solution and how to establish a trade-off between quality of the approximation and running time of the algorithms. Along these lines, the use of *Monte Carlo Algorithms* has been recently proposed $[3]$. Some counting problems have been shown in $[3]$ to admit of a Monte Carlo algorithm achieving a prefixed accuracy level in polynomial time. On the contrary, unless RP=NP, counting problems whose decision version is NP-complete cannot be solved, within a prefixed accuracy level, by a Monte Carlo algorithm having polynomial worst case time complexity, as it was shown in $[7]$. This theoretical result seems to contrast with the satisfactory behaviour exhibited in practice by simple Monte Carlo procedures: it is therefore interesting to evelute the running times *on the average* by means of a probabilistic analysis of the algorithms, assuming a reasonable stochastic model over the instances of the problem.

In Section 2, a general framework is established for a probabilistic analysis of the running time of Monte Carlo procedures, defining different levels of confidence in terms of stochastic convergence, as the instance size of the problem grows asymptotically large.

In Section 3, a probabilistic analysis is developed for a simple Monte Carlo procedure for the approximate solution of a #P-complete counting problem, the Intersection Cardinality Problem (ICP), whose decision version is NP-complete.

It is shown that, under mild assumptions on the parameters of the stochastic model, the running time of a Monte Carlo algorithm for the approximate solution of the ICP, achieving a prefixed accuracy level, is consistently less than its worst case time complexity.

2. A Framework for the Probabilistic Analysis of Monte Carlo Algorithms

A *Monte Carlo Algorithm* (MCA) for solving a counting problem π is based on a sampling procedure SP whose output X is a nonnegative unbiased estimator of the number M of feasible solutions. The sample mean

$$\bar{X} = \bar{X}(t) = \frac{1}{t} \sum_{i=1}^{t} X_i$$

obtained in a sequence of t independent trials is retained as an approximation to M, measuring its accuracy by means of the relative error $|\bar{X}-M|/M$, for M>0 (observe that, if M=0, unbiasedness and nonnegativity of X do imply $\bar{X}=0$).

An MCA is said to achieve an (ε,δ) *accuracy* if

$$\Pr \{|\bar{X}(t)-M|/M<\varepsilon| \quad M\geq 1\} \geq 1-\delta \tag{1}$$

for $\varepsilon>0$, $0<\delta<1$. The standard way of determining a priori the sample size t which guarantees the (ε,δ) accuracy of the MCA, is that of using Chebychev's inequality, thus leading to the choice of the value

$$t_o(n) = \frac{1}{\varepsilon^2 \delta} \sup_{I \in D_n} \{var(M)/M^2\} \qquad (2)$$

where $D_n = \{I \in \pi : |I| = n, M \neq 0\}$ is the class of instances of size n.

If T indicates the number of steps required by the sampling procedure, the overall time complexity of an (ε, δ)-MCA based on $t_o(n)$ independent trials is given by $0 (t_o(n) \cdot T)$.

Although for a few #P-hard counting problems it has been possible to design sampling strategies for which the time complexity $0(t_o(n) \cdot T)$ is polynomial in the input size $(\left[2\right], \left[3\right], \left[7\right])$, such a substantial computational saving with respect to exact algorithms cannot be achieved via (ε, δ) MCA's, unless RP=NP, for those counting problems whose decision version is NP-complete (as shown in $\left[7\right]$).

With respect to these latter problems, one could wonder whether the quoted negative result is due to the rather conservative attitude which leads to the "worst-case" choice of $t_o(n)$ or there are a relevant number of instances giving rise to values of $var(X)/M^2$ close to the supremum in (2). This question can be answered in a probabilistic framework, by assuming a reasonable distribution over D_n and by investigating the "average behaviour" of the expression (2) in terms of stochastic convergence, as n tends to infinity.

Specifically, let Pr^E denote the probability measure assumed over the set of instances D_n (E standing for "external"). Let also Pr^I (I standing for "internal") be the probability measure induced over the set of possible realizations of the output X by the random steps incorporated into the sampling procedure.

In order to investigate whether the a priori choice of the sample size $t_o(n)$ is too pessimistic, with respect to the assigned probability distribution over D_n, it is useful to define a threshold function $t_1(n)$ as follows, according to $\left[4\right]$:

$$\lim_{n \to \infty} \Pr^E \{t(n) > \frac{1}{\varepsilon^2 \delta} \frac{\text{var}(X)}{M^2} \quad |M \geq 1\} =$$

$$= \begin{cases} 1 & \text{iff} & \liminf_{n \to \infty} \dfrac{t(n)}{t_1(n)} > 1 \\\\ 0 & \text{iff} & \limsup_{n \to \infty} \dfrac{t(n)}{t_1(n)} < 1 \end{cases} \tag{3}$$

(Observe that var(X) is meant with respect to the internal probability measure).

The threshold function $t_1(n)$ can be interpreted as the minimum sample size which guarantees, through Chebychev's inequality, that (1) holds with external probability tending to one, as the input size n grows asymptotically large.

A third sensible choice for the sample size is given by

$$t_2(n) = E^E \{ \frac{1}{\varepsilon^2 \delta} \frac{\text{var}(X)}{M^2} \quad |M \geq 1\}.$$

It can be seen that $t_2(n)$ satisfies

$$E^E \{\Pr^I \{|\bar{X}(t_2(n)-M|/M < \varepsilon\} \ |M \geq 1\} > 1-\delta.$$

Notice that, by definition, the following relations hold among the functions $t_o(n)$, $t_1(n)$ and $t_2(n)$:

$$t_o(n) \geq t_2(n) \qquad \text{for any } n,$$

$$\liminf_{n \to \infty} \frac{t_o(n)}{t_1(n)} > 1 \ .$$

Moreover, observe that no similar relation between $t_1(n)$ and $t_2(n)$ can be established in general.

3. A probabilistic analysis of the ICP

In this section we are concerned with the following counting problem, indicated in the sequel as the *Intersection Cardinality Problem* (ICP). Given a collection of v finite sets D_1, D_2, ... D_v, each of cardinality at least two, and a collection $\mathcal{E} = \{ S_1, S_2, ...S_m \}$ of m subsets of the ground set $D = \overset{v}{\underset{i=1}{X}} D_i$, it is asked to determine the cardinality M of the intersection of the m subsets:

$$M = \bigcap_{i=1}^{m} S_i$$

As far as the representation of the subsets S_1, S_2, ...S_m is concerned, it will be assumed throughout the paper that they are described in *concise form*, instead of being explicitly listed, i.e. there exists an oracle ([8]) which says, within time polynomial in v, whether any point in D belongs to a subset $S_i \in \mathcal{E}$. Therefore, the size of an instance of the ICP is characterized by the input parameters m and v.

Several relevant counting problems are recognized to be particular cases of the ICP. For example, consider the problem of determining the number of feasible solutions of a 0,1-programming problem: in this case S_j (j=1,2,...m) is the set of feasible solutions to the j-th constraint (which, in fact, describes S_j in concise form).

A second problem (referred to as CNF-SAT) which can be interpreted as a particular case of the ICP, is that of counting the truth assignments to v binary variables satisfying a boolean formula in conjunctive normal form: each set S_j consists of all assignments verifying the j-th clause, which provides a concise definition of S_j. Observe that the cardinality n of the ground set D is exponential in the input parameter v; in the case of CNF-SAT, for example, $|D| = n = 2^v$.

From the point of view of computational complexity, the ICP can be easily seen to be #P-complete, since its restriction to CNF-SAT is known to be #P-complete. Also, the ICP is unlikely to admit of an MCA achieving (ε, δ)

accuracy in time polynomial in v and m, since its corresponding decision version is NP-complete. The simple MCA based on a Hit or Miss sampling strategy has a worst case running time which is exponential in the instance size.

Indeed, consider the following algorithm:

procedure HP (ICP)

    begin randomly choose a point d $\in$ D

$$\text{if} \quad d \in \bigcap_{j=1}^{m} S_i \quad \begin{array}{l} \text{then} \quad I(d)=1 \\ \text{else} \quad I(d)=0 \end{array}$$

    X = n·I(d)

    end .

It is easy to see that

$$E(X) = M$$

and

$$var(X) = M(n-M). \tag{4}$$

Hence, the MCA based on the sampling procedure HP(ICP) and achieving (ε, δ) accuracy, requires a worst case running time given by $O\left(\frac{1}{\varepsilon^2 \delta} \cdot m \cdot n \cdot t(HM)\right)$, where t(HM) is the number of steps necessary to test membership of a point d $\in$ D to a subset $S_j \in \mathcal{E}$. (Recall that t(HM) is polynomial in v).

The purpose of this section is that of investigating the probabilistic behaviour of the HM(ICP) procedure, assuming a natural *stochastic model* over the ICP instances of size v,m: for any point d $\in$ D and for any subset $S_j \in \mathcal{E}$, there is a probability p(v,m), depending from v and m, but independent from d and S_j, that d belongs to S_j:

$$Pr\{d \in S_j\} = p(v,m)$$

Under these assumptions, the distribution of the solution M can be determined :

Proposition 1

$$\Pr\{ M=h \mid M\geq 1\} = \frac{\binom{n}{h} \left[p(v,m)\right]^{mh} (1-\left[p(v,m)\right]^{m})^{n-h}}{1-(1-\left[p(v,m)\right]^{m})^{n}}$$

Proof : The proof follows from these facts:

(i) $\Pr\{d \in \bigcap_{i=1}^{m} S_j\} = \left[p(v,m)\right]^{m}$ for any $d \in D$;

(ii) $\Pr\{M=h\} = \binom{n}{h} \left[p(v,m)\right]^{mh} (1-\left[p(v,m)\right]^{m})^{n-h}$

(iii) $\Pr\{M=0\} = (1-\left[p(v,m)\right]^{m})^{n}$. ∎

Next, an upper bound on $t_1(m,v)$, with respect to the HP(ICP) procedure, is derived:

Theorem 1   If

$$\liminf_{v,m\to\infty} p(v,m)\, n^{1/m} > 1 \tag{5}$$

and if $t(v,m)$ is a function such that

$$\liminf_{v,m\to\infty} \frac{t(v,m)}{\dfrac{1}{\varepsilon^2\delta}\dfrac{1}{\left[p(v,m)\right]^{m}}} \geq 1 \tag{6}$$

then

$$\lim_{v,m\to\infty} \Pr^{E} \{\frac{1}{\varepsilon^2\delta} \frac{\mathrm{var}(X)}{M^2} < t(v,m) \mid M \geq 1\} = 1 \tag{7}$$

Proof   Let

$$\beta = \beta(v,m) = 1 - \frac{1}{\varepsilon^2\delta\cdot t(v,m) \left[p(v,m)\right]^{m}} ;$$

it is easy to check that relation (6) implies that $0<\beta<1$, for v,m large enough.

Since var(X) = $M(n-M)$, and due to the result established in Proposition 1, the following bound holds:

$$\Pr{}^{E} \{ \frac{1}{\varepsilon^2 \delta} \frac{\text{var}(X)}{M} > t(v,m) \mid M \geq 1 \} \leq$$

$$\leq \Pr{}^{E} \{ \frac{1}{\varepsilon^2 \delta} \frac{n}{M} > t(v,m) \mid M \geq 1 \}$$

$$= \Pr{}^{E} \{ M < \frac{n}{\varepsilon^2 \delta \, t(v,m)} \mid M \geq 1 \}$$

$$= \frac{\displaystyle\sum_{h=1}^{\left\lfloor \frac{n}{\varepsilon^2 \delta t(v,m)} \right\rfloor} \binom{n}{h} \left[p(v,m) \right]^{mh} (1 - \left[p(v,m) \right]^{m})^{n-h}}{1 - (1 - \left[p(v,m) \right]^{m})^{n}}$$

$$= \frac{\displaystyle\sum_{h=1}^{\left\lfloor (1-\beta) n \left[p(v,m) \right]^{m} \right\rfloor} \binom{n}{h} \left[p(v,m) \right]^{mh} (1 - \left[p(v,m) \right]^{m})^{n-h}}{1 - (1 - \left[p(v,m) \right]^{m})^{n}}$$

Applying Chernoff's inequality (see [5]) on the tails of the binomial distribution, the last expression is bounded from above by

$$\frac{\exp \{ -\beta^2 \, n \, \left[p(v,m) \right]^{m} / 2 \}}{1 - (1 - \left[p(v,m) \right]^{m})^{n}} \tag{8}$$

Because of condition (5) the numerator of expression (8) tends to zero; in fact

$$\lim_{v,m \to \infty} -\beta^2 \, n \, \left[p(v,m) \right]^{m} = -\infty \quad .$$

The denominator in (8) can be shown to be far from zero, whenever the input grows asymptotically large.

In fact, by condition (5)

$$\liminf_{v,m\to\infty} \left(1- \left[p(v,m)\right]^{-m}\right)^n \leq \liminf_{v,m\to\infty} \left(1- \frac{1}{n}\right)^n = e^{-1}$$

and hence

$$\lim_{v,m\to\infty} 1-\left(1- \left[p(v,m)\right]^{m}\right)^n \geq 1-e^{-1} > 0.$$

In conclusion, expression (8) tends to zero, if $v,m\to\infty$ and the theorem is proved. ∎

Notice that condition (5) is rather mild. Indeed, suppose, as in $\left[6\right]$ for CNF-SAT, that $m=\alpha v$, with $\alpha \in \mathbb{R}^+$, and p is constant; then (5) is satisfied for every (α,p) lying in the shaded region of figure 1. Observe also that if $v<m$ and $p = \frac{1}{2}$, then (5) holds.

As a consequence of the definition of $t_1(v,m)$ and Theorem 1, if (5) is true, then the following bound from above on $t_1(v,m)$ is established for large v and m:

$$\limsup_{v,m\to\infty} \frac{t_1(v,m)}{\frac{1}{\varepsilon^2\delta} \frac{1}{\left[p(v,m)\right]^m}} \leq 1 . \tag{9}$$

The following combinatorial lemmas will be useful in the sequel: essentially they describe a lower and an upper bound on the expectation of the inverse of a truncated binomial random variable.

Lemma 1   If $0<p<1$, then

$$\sum_{k=1}^{n} \frac{1}{k} \binom{n}{k} p^k (1-p)^{n-k} \leq \frac{2}{p(n+1)} .$$

Proof

$$\sum_{k=1}^{n} \frac{1}{k} \binom{n}{k} p^k (1-p)^{n-k} =$$

$$= \sum_{k=1}^{n} \frac{1}{k} \frac{n+1}{k+1} \frac{k+1}{n+1} \frac{n!}{k!\,(n+k)!} p^k (1-p)^{n-k}$$

$$= \sum_{k=1}^{n} \frac{1}{k} \frac{(n+1)!}{(k+1)!\,(n-k)!} \frac{k+1}{n+1} \frac{p^{k+1}}{p} (1-p)^{n-k}$$

$$= \frac{1}{p(n+1)} \sum_{k=1}^{n} (1+\frac{1}{k}) \binom{n+1}{k+1} p^{k+1} (1-p)^{n-k}$$

$$\leq \frac{2}{p(n+1)} \sum_{k=1}^{n} \binom{n+1}{k+1} p^{k+1} (1-p)^{n-k}$$

$$\leq \frac{2}{p(n+1)} \sum_{k=0}^{n} \binom{n+1}{k} p^k (1-p)^{n-k+1}$$

$$= \frac{2}{p(n+1)} \quad \blacksquare$$

Lemma 2   It $0<p<1$, then

$$\frac{\sum_{k=1}^{n} \frac{1}{k} \binom{n}{k} p^k (1-p)^{n-k}}{1-(1-p)^n} \geq \frac{1-(1-p)^n}{n\,p} .$$

Proof: Applying Cauchy-Schwarz inequality to the functions $\sqrt{k}$ and $\dfrac{1}{\sqrt{k}}$, with the measure given by the truncated binomial distribution with parameters n and p, one obtains:

$$\left\{ \sum_{k=1}^{n} \frac{\binom{n}{k} \, p^k (1-p)^{n-k}}{1-(1-p)^k} \right\}^2 \leq$$

$$\sum_{k=1}^{n} \frac{k \binom{n}{k} \, p^k (1-p)^{n-k}}{1-(1-p)^n} \quad \sum_{k=1}^{n} \frac{\frac{1}{k} \binom{n}{k} \, p^k (1-p)^{n-k}}{1-(1-p)^n} \quad .$$

The left handside is equal to one, while the first factor of the right handside is the expectation of a truncated binomial random variable. Therefore

$$1 \leq \frac{np}{1-(1-p)^n} \sum_{k=1}^{n} \frac{\frac{1}{k} \binom{n}{k} \, p^k (1-p)^{n-k}}{1-(1-p)^n} \quad . \qquad \blacksquare$$

In Section 2 we pointed out that no relation can be stated in general terms between $t_1(n)$ and $t_2(n)$; however, under the stochastic assumptions made in this Section it is possible to prove the following result:

Theorem 2   If $t(v,m)$ is a function such that

$$\liminf_{v,m \to \infty} \frac{t(v,m)}{t_2(v,m)} \geq 1$$

then

$$\lim_{v,m \to \infty} \Pr^E \{ \frac{1}{\varepsilon^2 \delta} \frac{\text{var}(X)}{M^2} < t(v,m) \mid M \geq 1 \} = 0$$

Proof

$$\lim_{v,m \to \infty} \Pr^E \{ \frac{1}{\varepsilon^2 \delta} \frac{\text{var}(X)}{M^2} < t(v,m) \mid M \geq 1 \} \leq$$

$$\leq \lim_{v,m \to \infty} \Pr^E \{ \frac{1}{\varepsilon^2 \delta} \frac{\text{var}(X)}{M^2} < t_2(v,m) \mid M \geq 1 \}$$

$$\leq \lim_{v,m \to \infty} \Pr^E \{ \frac{1}{M} < E^E \left[\frac{1}{M} \mid M \geq 1 \right] \mid M \geq 1 \} \quad .$$

$$\leq \lim_{v,m\to\infty} \Pr^E \{\frac{1}{M} > \frac{2}{\left[p(v,m)\right]^m} \frac{1}{n+1} \frac{1}{1-(1-\left[p(v,m)\right]^m)^n} \mid M \geq 1\}$$

(the last inequality is a consequence of Proposition 1 and Lemma 1)

$$= \lim_{v,m\to\infty} \Pr^E \{M \geq \left[\frac{1}{2}\left[p(v,m)\right]^m(n+1)\left[1-(1-\left[p(v,m)\right]^m)^n\right]\right] \mid M \geq 1\}$$

(since:

$$0 < \frac{1}{2}\left[p(v,m)\right]^m \{1-(1-\left[p(v,m)\right]^m)^n\} \quad < 1 \quad)$$

$$\leq \lim_{v,m\to\infty} \Pr^E \{M \geq n+1 \mid M \geq 1\} \quad = 0$$

being $M \leq n$. ■

Again, it follows by definition that

$$\limsup_{v,m} \frac{t_2(v,m)}{t_1(v,m)} < 1 \tag{10}$$

In the next theorem a lower bound on $t_2(v,m)$ is derived.

Theorem 3

$$t_2(v,m) \geq \frac{1}{\varepsilon^2\delta} \frac{1-(1-\left[p(v,m)\right]^m)^n}{\left[p(v,m)\right]^m}$$

Proof: By definition and by equality (4)

$$t_2(v,m) = \frac{n}{\varepsilon^2\delta} E^E \left[\frac{1}{M}\mid M \geq 1\right] - \frac{1}{\varepsilon^2\delta} \quad ;$$

applying Lemma 2 and Proposition 1, the result easily follows. ■

By relation (10) and by Theorem 3 it follows that

$$1 > \limsup_{v,m\to\infty} \frac{t_2(v,m)}{t_1(v,m)} \geq \limsup_{v,m\to\infty} \frac{1}{\varepsilon^2\delta t_1(v,m)} \frac{1-(1-\left[p(v,m)\right]^m)^n}{\left[p(v,m)\right]^m} \quad ;$$

therefore

$$\lim_{v,m \to \infty} \inf \frac{t_1(v,m)}{\frac{1}{\varepsilon^2 \delta} \frac{1-(1-\left[p(v,m)\right]^m)^n}{\left[p(v,m)\right]^m}} > 1$$

and by condition (5)

$$\lim_{v,m \to \infty} \inf \frac{t_1(v,m)}{\frac{1}{\varepsilon^2 \delta} \frac{1-e^{-1}}{\left[p(v,m)\right]^m}} > 1 \tag{12}$$

Hence, in light of (9)

$$\lim_{v,m \to \infty} \frac{t_1(v,m)}{\frac{1}{\varepsilon^2 \delta} \frac{1}{\left[p(v,m)\right]^m}} = 1,$$

and this characterizes the asymptotical behaviour of the threshold $t_1(v,m)$, whenever condition (5) is satisfied.

In conclusion, the MCA based on the HM(ICP) procedure achieves (ε,δ)-accuracy, with (external) probability tending to one, in a number of steps given by

$$O\left(\frac{1}{\varepsilon^2 \delta} \frac{m}{\left[p(v,m)\right]^m} \ T \ (HM)\right)$$

which, under assumption (5), is less than its worst case running time. In particular, the computational saving becomes more consistent whenever $p(v,m)$ grows faster than $n^{-\frac{1}{m}}$.

This result shows that even in the case of a simple Monte Carlo algorithm and under mild assumptions the number of elementary steps required to achieve (ε,δ)-accuracy on the average is substantially less than in the worst-case.

References

[1] M.O. Ball, "Complexity of Network Reliability Computation", Networks,
     10 (1980), 153-165.

[2] R.M. Karp, M.G. Luby, "A Monte Carlo Method for Estimating the Failure
     Probability of an n-Component System", T.R. UCB/CSD 83/17 Univ. of
     California, Berkeley (1983).

[3] R.M. Karp, M.G. Luby, "Monte Carlo Algorithms for Enumeration and
     Reliability Problems", T.R. Univ. of California, Berkeley (1984).

[4] P. Erdös, A. Renyi, "On the Evolution of Random Graphs", Math. Kutato
     Int. Kozl., 5 (1960),17-60.

[5] P. Erdös, J. Spencer, "Probabilistic Methods in Combinatorics", Aca-
     demic Press, New York, 1974 .

[6] J. Franco, M. Paull, "Probabilistic Analysis of the Davis Putnam
     Procedure for Solving the Satisfability Problem", DAM, 5 (1983), 77-87.

[7] A. Frigessi, C. Vercellis, "An Analysis of Monte Carlo Algorithms for
     Counting Problems", Calcolo, to appear.

[8] M.R. Garey, D.S. Johnson, "Computers and Intractability: a Guide to
     the Theory of NP-Completeness", Freeman, San Francisco,1979.

[9] J.S. Provan, M.O. Ball, "The Complexity of Counting Cuts and of
     Computing the Probability that a Graph is Connected", SIAM J. Comp.,12
     (1983), 777-788.

[10] L.G. Valiant, "The Complexity of Computing the Permanent", Theor. Comp.
     Sci., 8 (1979), 189-201.

[11] L.G. Valiant, "The Complexity of Enumeration and Reliability Problems",
     SIAM J. Comp., 8 (1979), 410-421.

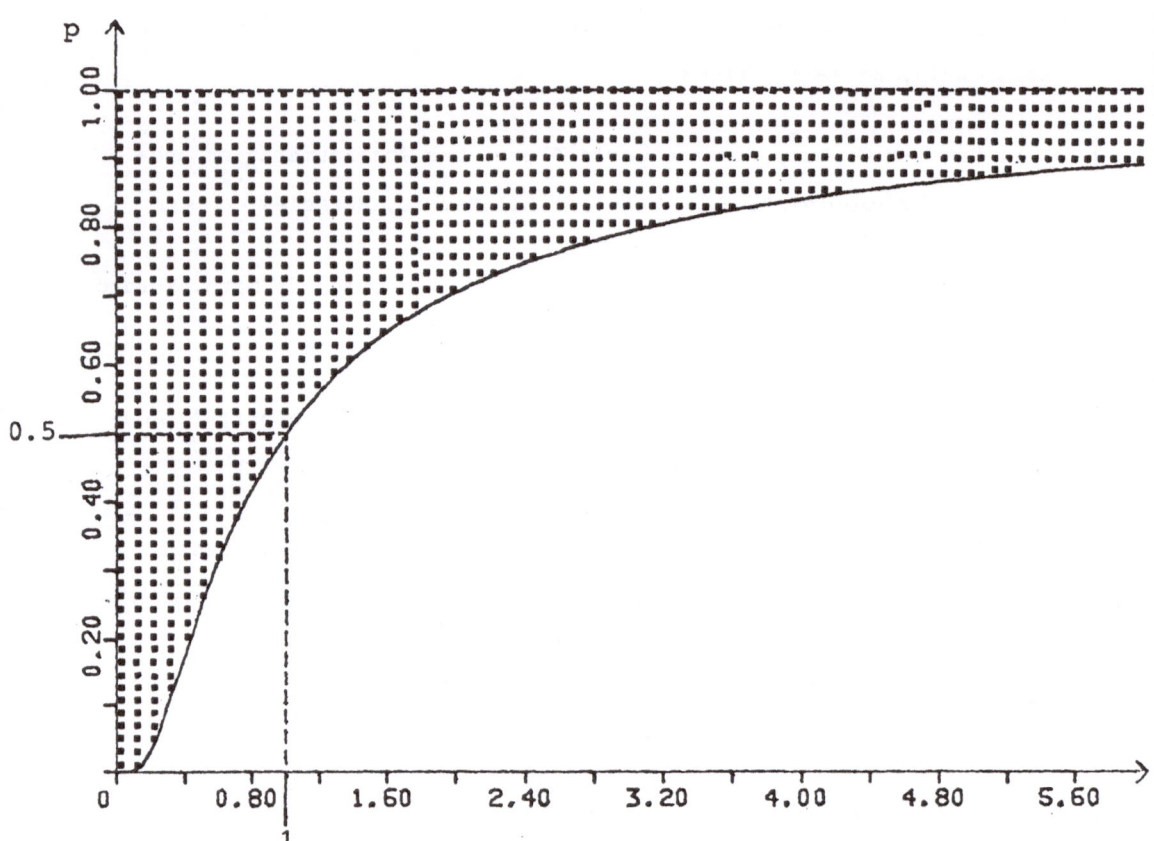

Figure 1

AN ALGORITHM FOR SOLVING LINEAR RANDOM
DIFFERENTIAL AND INTEGRAL EQUATIONS

Melvin D. Lax

Department of Mathematics and Computer Science
California State University, Long Beach
Long Beach, California, USA 90840

I. INTRODUCTION

The crucial importance of deterministic integral equations and differential equations to modelling phenomena in science and engineering has been
long an established fact. However, the known functions used in such models
often are not known exactly, but instead are subject to uncertainty (e.g.
errors in measurement or variation of physical properties in samples used
to provide known quantities.) To account for these uncertainties applied
mathematicians have turned increasingly in the last 25 years to stochastic
models in which the deterministic integral or differential equation is replaced by a random integral or differential equation. This has led to
successful applications to numerous areas including control systems, communication theory, mechanical vibrations, chemical kinetics, and turbulence
theory. (Several interesting applications can be found in Tsokos and
Padgett [4].)

The major drawback to the use of random integral and differential
equations has been the increased difficulty in solving them as compared to
their deterministic counterparts. As closed form solutions are usually
not attainable, research in recent years has focused on developing methods
to approximate solutions. Perturbation techniques, hierarchy techniques,
stochastic Green's function method, reduction to deterministic partial
differential equations, finite element methods, successive approximation,
stochastic approximation, and the method of moments all have been applied

successfully to certain classes of random equations. In particular, the method of moments (see Vorobyev's monograph [5] for a complete derivation and application to deterministic problems) has been proved to provide approximations that converge to the solution of a class of random integral equations and random ordinary differential equations in previous papers by the author [1,2,3]. The purpose of this paper is to present an algorithm for using the method of moments to generate numerical approximations of random integral equations, to discuss its extensions (particularly to random ordinary differential equations), and to give some examples illustrating its effectiveness.

2. THE ALGORITHM

Consider the random linear Volterra integral equation of the second kind

(1)
$$X(t,\omega) = \int_0^t K(t,s,\omega)\, X\,(s,\omega)\,ds + V(t,\omega)$$

where ω is an element of a probability space Ω, t is real, $V(t,\omega)$ is a bounded stochastic process with square integrable sample functions which is smooth in the sense that there exists a sequence $\{V_m(t,\omega)\}$ such that each V_m is a discrete stochastic process whose m sample functions are also sample functions of V and

$$\lim_{m\to\infty} \int_0^1 E\{[\,V(t,\omega) - V_m(t,\omega)]^2\}dt = 0,$$

and where $K(t,s,\omega)$ is a stochastic process satisfying analogous conditions to those just listed for $V(t,\omega)$. Then (see [3]) the method of moments approximate solutions $X_k(t,\omega)$ of (1) converge to the solution $X(t,\omega)$ of (1) in the sense that

$$\lim_{k\to\infty} \int_0^1 E\{[\,X(t,\omega) - X_k(t,\omega)]^2\}dt = 0.$$

We may apply the following algorithm.

1. Choose an initial basis function z_0. (While this choice is arbitrary, the best results usually are obtained by choosing z_0 to be $V(t,\omega)$ or the first few terms of a random Taylor series or a random Fourier series for $V(t,\omega)$.)

2. Compute
$$z_i = \int_0^t K(t,s,\omega)\, z_{i-1}(s,\omega)\,ds, \quad i = 1, \ldots, n.$$

    (One may numerically integrate to get the z_i's at just those values of t needed to complete steps 3 and 6. Alternatively one may attempt to approximate z_0 and $K(t,s,\omega)$ by truncated random Taylor series or truncated random Fourier series. In many cases the method of moments converges extremely fast so that even if n is chosen to be small (say 4 or 5), several digits accuracy will be obtained.)

3. Compute
$$b_{ij} = \int_0^1 E\{z_i(t,\omega)\, z_j(t,\omega)\}\,dt, \quad i = 0, \ldots, n-1 \quad \text{and}$$

    $j = 0, \ldots, n.$ (Numerical integration should be used here.)

4. Solve the systems of equations
$$b_{00}c_0 \;+\; b_{01}c_1 \;+\; \cdots \;+\; b_{0,n-1}c_{n-1} \;=\; -\,b_{0n}$$
$$b_{10}c_0 \;+\; b_{11}c_1 \;+\; \cdots \;+\; b_{1,n-1}c_{n-1} \;=\; -\,b_{1n}$$
$$\vdots$$
$$b_{n-1,0}c_0 \;+\; b_{n-1,1}c_1 \;+\; \cdots \;+\; b_{n-1,n-1}c_{n-1} \;=\; -\,b_{n-1,n}\;.$$

    (This should be done using a very accurate numerical method as this system can be ill-conditioned. Gaussian elimination with complete pivoting done in double precision has proved successful.)

5. Compute
$$k_0 = 1 - \frac{c_0}{1 + \sum_{j=0}^{n-1} c_j}$$

$$k_i = k_{i-1} - \frac{c_i}{1 + \sum_{j=0}^{n-1} c_j}, \quad i = 1, \ldots, n-1$$

6. The approximate solution of (1) is $X_n(t,\omega) = \sum_{i=0}^{n-1} k_i z_i(t,\omega)$. The

approximate mean $E\{X_n(t_i,\omega)\}$ and the approximate autocorrelation

$E\{X_n(t_i,\omega) X_n(t_j,\omega)\}$ can be computed now at the desired values of

t_i, t_j, $i = 1, \ldots, m$, $j = 1, \ldots, \ell$, in the interval $[0,1]$.

To gauge the accuracy of the results found, repeat steps 1-6
with n replaced by $n + 1$. The results should be accurate to at
least as many digits as are the same in the two iterations.

3. EXTENSIONS.

The algorithm can be extended in the following ways:

1. Clearly the solution of the Volterra equation (1) can be computed at any
    finite value of t by using an appropriate change of variable before
    applying the algorithm.

2. Under more restrictive conditions [1] it has been shown that the
    method of moments yields convergent approximate solutions of the random
    Fredholm integral equation of the same form as equation (1) except that
    the t in the upper limit of integration is replaced by a known con-
    stant c. The algorithm will work for such equations if the z_i's in
    step 2 are computed as follows:

$$z_i = \int_0^c K(t,s,\omega) z_{i-1}(s,\omega) ds .$$

Care should be taken to be sure that 1 is not an eigenvalue or the
limit of a sequence of eigenvalues of the random Fredholm equation.

3. The random initial value problem

    (2) $Y^{(n)} + Q_1(t,\omega) Y^{(n-1)} + \ldots + Q_n(t,\omega) Y = H(t,\omega)$,

    $Y(0) = Y''(0) = \ldots = Y^{(n-1)}(0) = 0$, can be written in the form
    of equation (1) with $X(t,\omega) = Y^{(n)}(t,\omega)$, $V(t,\omega) = H(t,\omega)$, and

$$K(t,s,\omega) = - \sum_{j=1}^{n} Q_j(t,\omega) \frac{(t-s)^{j-1}}{(j-1)!} \, .$$

Then the algorithm may be used to solve for the mean and autocorrelation of $Y^{(n)}(t,\omega)$. These results may be integrated n times to yield the mean and autocorrelation of $Y(t,\omega)$. Of course, if the original initial value problem has nonhomogeneous initial values a simple transformation can be applied to put it in the form of initial value problem (2).

4. In solving two point random boundary value problems, one might use the associated random Green's function to transform the problem into a random Fredholm integral equation and then proceed as discussed in extension 2. However, in many cases a more expeditious approach is found by using a different transformation. For example, consider

(3) $Y'' + Q(t,\omega) Y = H(t,\omega) , \qquad Y(0) = Y(1) = 0.$

If $G(t,s)$ is the (deterministic) Green's function associated with the problem $y'' = 0$, $y(0) = y(1) = 0$, then it follows that problem (3) can be written as

$$Y(t,\omega) = \int_0^1 - G(t,s) Q(s,\omega) Y(s,\omega) ds + Z(t,\omega)$$

where $$Z(t,\omega) = \int_0^1 G(t,s) H(s,\omega) ds.$$

Now the algorithm can be used after modifying it so that step 2 has

$$z_i = \int_0^1 - G(t,s) Q(s,\omega) z_{i-1}(s,\omega) ds \quad ,$$

and step 3 has

$$b_{ij} = \int_0^1 - E\{z_i''(t,\omega) z_j(t,\omega)\} dt.$$

Similar transformations will be effective for other two point random boundary value problems [2]. Again, as in extension 2, care should be taken to stay away from eigenvalues.

4. EXAMPLES.

In order to gauge accuracy and speed of convergence, the algorithm has been applied to the following examples whose exact solutions are known.

1. $Y'' + HY = V_1 + V_2t$, $Y(0) = Y'(0) = 0$ where H, V_1, V_2 are random variables; H is independent of V_1, V_2; H is uniformly distributed on $[2,4]$; and $E\{V_1\} = 1$, $E\{V_2\} = 2$, $E\{V_1^2\} = 12$, $E\{V_1V_2\} = 4$, $E\{V_2^2\} = 8$.

This random initial value problem is converted to the random Volterra integral equation

$$X(t) = \int_0^t (s - t)HX(s)ds + V_1 + V_2t$$

where $X(t,\omega) = Y''(t,\omega)$.

The algorithm is applied and then the solution is integrated twice.
The following results are obtained with $n = 6$.

TABLE I

Numerical results for the mean of the solution of example 1

t	Method of moments solution	Exact solution	Error
0.2	0.02245153888	0.02245153891	-0.0000000003
0.4	0.09767995740	0.09767995743	-0.0000000003
0.6	0.23260646867	0.23260646870	-0.0000000003
0.8	0.42711378570	0.42711378576	-0.0000000006
1.0	0.67412212116	0.67412212124	-0.0000000008

2. $$X(t) = \int_0^1 (H_1 + H_2 ts) X(s) ds + V$$

where H_1, H_2, V are independent random variables, H_1 is uniformly distributed on [2,4], H_2 is uniformly distributed on ⌈1,2⌉, and $E\{V\} = 1$, $E\{V^2\} = 12$. The following results are obtained with $n = 5$.

TABLE 2

Numerical results for the autocorrelation of the solution of example 2

t_1	t_2	Method of moments solution	Exact solution	Error
0.2	0.2	0.00492888803	0.00492888804	-0.00000000001
0.2	0.6	0.04299721310	0.04299721315	-0.00000000005
0.2	1.0	0.10695777444	0.10695777457	-0.00000000013
0.6	0.6	0.37857047006	0.37857047011	-0.00000000005
0.6	1.0	0.95098373736	0.95098373747	-0.00000000011
1.0	1.0	2.41388753833	2.41388753860	-0.00000000027

TABLE 3

Numerical results for the mean of the solution of example 2

t	Method of moments solution	Exact solution	Error
0.0	-0.10937	-0.11029	0.00092
0.2	-0.14292	-0.14348	0.00056
0.4	-0.17647	-0.17667	0.00020
0.6	-0.21002	-0.20987	-0.00015
0.8	-0.24358	-0.24306	-0.00052
1.0	-0.27713	-0.27625	-0.00088

Results of similar accuracy were obtained for the autocorrelation.

3. $-Y'' + RY = V \sin \pi t$, $Y(0) = Y(1) = 0$ where R,V are independent random variables, R is uniformly distributed on [-4,-2], and $E\{V\} = 1$, $E\{V^2\} = 12$. The following results are obtained with $n = 3$.

TABLE 4

Numerical results for the mean of the solution of example 3

t	Method of moments solution	Exact solution	Error
0.2	0.08617534	0.08617535	-0.00000001
0.4	0.13943463	0.13943466	-0.00000003
0.6	0.13943463	0.13943466	-0.00000003
0.8	0.08617536	0.08617538	-0.00000002

Results of similar accuracy were obtained for the autocorrelation.

5. SUMMARY.

The algorithm is an effective, inexpensive way to generate approximate solutions to random integral and differential equations. There
are n integrations in step 2, $\frac{1}{2}n^2 + \frac{3}{2}n + 1$ integrations in step 3,
an n x n system of equations is solved in step 4, and there are some
multiplications and additions in step 5 whose number depend on how many
values of t are used. Thus few function evaluations are required for
small n; and small n should be sufficient due to the rapid convergence of the method of moments (see [5,p.36]). Indeed all the examples of section 4 were done with less than 90 seconds of computer
execution time. Table 5 demonstrates how the accuracy of the approximate
solutions of the examples increases as n, the number of basis functions
(z_i's), increases from 2 to 6.

Experience has shown that the approximate solutions of random Fredholm
integral equations tend to converge somewhat slower. Also close proximity to eigenvalues can dramatically slow convergence for both random
Fredholm integral equations and random boundary value problems. Never-
theless the algorithm and its extensions provide an excellent procedure
for finding accurate approximations of the mean and autocorrelation of
the solutions of random integral equations and random differential
equations.

TABLE 5

The number of accurate digits in the approximate solutions generated by the algorithm

Example				n		
		2	3	4	5	6
1	mean	1	3	5	7	9
	autocorrelation	1	3	4	6	9
2	mean	0	0	2	3	3
	autocorrelation	0	0	2	2	2
3	mean	4	6	8	8	8
	autocorrelation	3	5	8	8	8

6. REFERENCES.

   1. M.D. Lax, Method of Moments Approximate Solutions of Random Linear Integral Equations, J. Math. Anal. Appl. 58 (1977), 46-55.

   2. M.D. Lax, Obtaining Approximate Solutions of Random Differential Equations by means of the Method of Moments, Approximate Solution of Random Equations, A.T. Bharucha-Reid, ed., North-Holland, New York, 1979.

   3. M.D. Lax, Solving Random Linear Volterra Integral Equations Using the Method of Moments, J. Integral Eqns. 3 (1981), 357-363.

   4. C.P. Tsokos and W.J. Padgett, Random Integral Equations with Applications to Life Sciences and Engineering, Academic Press, New York, 1974.

   5. Yu. V. Vorobyev, Method of Moments in Applied Mathematics, Gordon and Breach, New York, 1965.

GROWTH VERSUS SECURITY IN A RISKY INVESTMENT MODEL

L.C. MacLean   and   W.T. Ziemba
Dalhousie University   University of British Columbia

1. INTRODUCTION

We consider a risky investment model where a decision maker is presented at each
point in time with the problem of wagering a portion of his current capital on a
sequence of investments with uncertain outcome. The goal of the investor is simply
accumulation of capital. Even when the investments have positive expected return
the problem is nontrivial. If the decision maker invests all his fortune (thus maximizing the expected value of his fortune) he will eventually go broke. It has been
suggested that the decision maker seek to maximize the expected value of the log of
his fortune [7]. The resulting "Kelly" strategy has many useful properties. In
particular it maximizes the rate of growth of capital. Furthermore, the probability
of eventually going broke is zero. However, it is possible to experience substantial
short-run losses.

In this paper we consider the issues of growth and security in risky investment
models. Some of the pertinent properties of the optimal growth (Kelly) strategy are
reviewed in section 3. Strategies offering greater security are considered in section 4. Then in Section 5 we consider the trade off between growth and security.

2. INVESTMENT MODEL

Suppose we have n investment opportunities and starting with initial capital
F_0 we are considering the strategy $\gamma_t = (\gamma_{1t}, \ldots, \gamma_{nt})$ at time t , where γ_{it} is
the fraction of current capital invested in opportunity i . As well we have a pro-
bability space (Ω, B, P) and the function $\phi : \Omega \times R^n \rightarrow R$ representing the return
on the capital invested. So given the outcome $\omega \in \Omega$ and the strategy $\gamma \in R^n$ we
have the return $\phi(\omega, \gamma F_0)$. We will assume that the return function ϕ is **homo-
geneous** $(\phi(\omega, \gamma F_0) = F_0 \phi(\omega, \gamma))$ and **favorable** (for at least one i , $E_\omega \phi(\omega, e_i) > 0$,
where $e_i = (\gamma_1, \ldots, \gamma_n)$, $\gamma_i = 1$, $\gamma_j = 0$ for $j \neq i$). Under these assumptions
we have the capital at time T given by

$$F_T(\omega^T; \gamma^T) = F_0 \, \Pi_{t-1}^{T}(1 + \phi(\omega_t, \gamma_t)) \quad , \tag{2.1}$$

where $\omega^T = (\omega_1, \ldots, \omega_T)$ and $\gamma^T = (\gamma_1, \ldots, \gamma_T)$.

Consider the strategy set $D = \{(\gamma_1,...,\gamma_t,...)|\gamma_t \in R^n, \forall t\}$. Then the object-
ive for this problem is to choose an investment strategy $\gamma^\infty \in D$ so that the accumulation of capital is somehow optimal. A much discussed approach to the problem is to

$$\text{maximize } \{\lim_{T\to\infty} E \log[F_T(\omega^T;\gamma^T)]^{\frac{1}{T}}\} \quad . \tag{2.2}$$

There are many reasons why this is a useful formulation of the objective. In particular, if we rewrite (2.1) as

$$F_T(\omega^T;\gamma^T) = F_0 \exp[\textstyle\sum_{t=1}^{T} \log(1 + \phi(\omega_t,\gamma_t))] \tag{2.3}$$

and consider the average growth rate along any path as $\frac{1}{T}\sum_{t=1}^{T} \log(1 + \phi(\omega_t,\gamma_t)) = $
$G_T(\omega^T;\gamma^T)$, then from the law of large numbers $G_T(\omega^T;\gamma^T)$ converges to

$$G(\bar{\gamma}) = E_\omega \log(1 + \phi(\omega,\bar{\gamma})) \quad , \tag{2.4}$$

where $\bar{\gamma} \in R^n$.

So (2.2) reduces to finding the one period strategy which maximizes the mean
growth rate. Most of the literature on this problem deals with properties of this
optimal fixed fraction (Kelly) strategy. We will review these results in the following section, but first we present some applications of the model.

Example 2.1: Blackjack

If $\Omega = \{0,1\}$ and $\phi(0,\gamma) = \gamma$ with probability p , and $\phi(1,\gamma) = -\gamma$ with probability $q = 1 - p$, then the model fits the classic game of blackjack (or "21") where
the bettor wins an amount equal to his bet with probability p or loses his bet with
probability q . In this case the mean growth rate becomes $E \log(1 + \phi(\omega,\gamma)) = $
$p \log(1 + \gamma) + q \log(1 - \gamma)$. Simple calculus gives the optimal fixed fraction strategy $\gamma^* = p - q$ if $E\phi > 0$; $\gamma^* = 0$ if $E\phi \le 0$. Since in fact blackjack is a
favorable game with $E\phi = .51 - .49 = .02$, the Kelly strategy would wager 2% of the
fortune at each play of the game.

Example 2.2: Horseracing

Suppose we have n horses entered in a race. Of the n final positions at the
finish of the race only the first three (win, place, show) have positive return to
the bettor, so for the remaining positions you lose the amount of your bet. Then
$\Omega = \{1,2,3),...,(i,j,k),...,(n-2,n-1,n)\}$ the set of all (win, place, show) outcomes
with probability p_{ijk} . The action we are considering is wagering the fractions
$\gamma_{i1}, \gamma_{i2}, \gamma_{i3}$ of our fortune F_0 on horse i to win, place or show respectively.
We denote by γ the $n \times 3$ matrix of wager fractions, where $\sum_{i=1}^{n} \sum_{j=1}^{3} \gamma_{ij} \le 1$.

The return function for a particular (i,j,k) outcome is a rather complicated expression[1] given by

$$\phi((i,j,k),\gamma) = (QW - W_i)\frac{\gamma_{i1}}{W_i} + (QP - P_i - P_j)(\frac{\gamma_{i2}}{P_i} + \frac{\gamma_{j2}}{P_j})$$

$$+ (QS - S_i - S_j - S_k)(\frac{\gamma_{i3}}{S_i} + \frac{\gamma_{j3}}{S_j} + \frac{\gamma_{k3}}{S_k}) \tag{2.5}$$

$$- (\sum_{\ell \neq i} \gamma_{\ell 1} + \sum_{\ell \neq i,j} \gamma_{\ell 2} + \sum_{\ell \neq i,j,k} \gamma_{\ell 3}) \quad ,$$

where $Q = 1$ - the track take, W_i, P_j and S_k are the total amounts bet to win, place and show on the indicated horses, respectively, and $W = \sum W_i$, $P = \sum W_j$, $S = \sum S_k$. The optimal strategy γ^* is found by solving the problem

$$\text{maximize } \{\sum_i \sum_{j \neq i} \sum_{k \neq i,j} P_{ijk} \log(1 + \phi((i,j,k),\gamma)) | \sum \sum \gamma_{ij} \leq 1, \gamma_{ij} \geq 0\} \quad .$$

3. OPTIMAL GROWTH

The Kelly strategy introduced in section 2 is optimal in the sense that it maximizes the mean growth rate of capital. However, there are many other important propties of this strategy and we will review them in this section. The results focus on two concepts: capital growth and net return on investment.

3.1 Capital Growth

Consider the set of fixed fraction strategies

$$D^* = \{(\gamma_1,\ldots,\gamma_t,\ldots) | \gamma_t = \gamma \in R^n, t\} \quad . \tag{3.1}$$

For $\gamma \in D^*$ we have the capital at time T given by $F_T(\omega^T,\gamma) = F_0 \Pi_{t=1}^T (1 + \phi(\omega_t,\gamma))$. Then the Kelly strategy is the fixed fraction strategy which maximizes $E(\log F_T)$.

Theorem 3.1 (Finkelstein & Whitley)

There exists $\bar{\gamma} \in D^*$ such that $E(\log F_T(\omega^T,\bar{\gamma})) = \max\{E(\log F_T(\omega^T,\gamma)) | \gamma \in D^*\}$. Furthermore $E(\log F_T(\omega^T,\bar{\gamma})) = \max\{E(\log F_T(\omega^T,\gamma)) | \gamma \in D^*\}$ if and only if $G(\bar{\gamma}) = \max\{G(\gamma) | \gamma \in D^*\}$.

---

[1] The return function (2.4) is a simplification of a function in Hausch (1981). In making the simplification it is assumed that our bets are small relative to the total amount bet, that is, we do not affect the odds.

So for the possibly more appealing criterion of expected log of wealth in any period, the Kelly strategy is the best among fixed fraction strategies.

If we define $T(\omega^{\infty}, \gamma^{\infty}, x)$ for $\gamma^{\infty} \in D$, $x > F_0$, as

$$T(\omega^{\infty}, \gamma^{\infty}; x) = \{\text{smallest } t \text{ such that } F_t(\omega^{\infty}, \gamma^{\infty}) \geq x\} \qquad (3.2)$$

then we have the following result.

Theorem 3.2 (Breiman)

If $\log (1 + \phi(\omega, \bar{\gamma}))$ is a non lattice random variable, then we have

$$\lim_{x \to \infty} [ET(\omega^{\infty}, \gamma^{\infty}, x) - ET(\omega^{\infty}, \bar{\gamma}, x)] \leq 0 \quad .$$

Again the Kelly strategy is optimal among all strategies in the sense that the expected time to reach large goals is least with that strategy.

If we look at the relative levels of wealth for Kelly and other strategies we have

Theorem 3.3 (Finkelstein & Whitley)

$F_T/\bar{F}_T = F_T(\omega^{\infty}, \gamma^{\infty})/F_T(\omega^{\infty}, \bar{\gamma})$ is a supermartingale with $E(F_T/\bar{F}_T) \leq 1$ and $E(\lim F_T/\bar{F}_T) \leq 1$.

This result addresses the quantity of most interest-accumulated capital -- but the expectation of the ratio of random variables is difficult to interpret. Indeed it is easy to construct strategies where the wealth ratio is exactly 1 .

Theorem 3.4 (Finkelstein & Whitley)

Suppose γ^{∞} is a strategy for which $\gamma_{ti} = 0$ if $\bar{\gamma}_i = 0$, $i = 1,\ldots,n$, for all t . Also if $\sum_{i=1}^{n} \bar{\gamma}_i = 1$ then $\sum_{i=1}^{n} \gamma_{ti} = 1$ for all t . Then $F_T/\bar{F}_T$ is a martingale with $E(F_T/\bar{F}_T) = 1$.

We see, therefore, that investing in the same options as the Kelly strategy leads to an equivalent expected wealth ratio, regardless of the proportions invested in those options.

3.2 Return on Investment

Our discussion of investment strategies to this point has focused on the growth of capital. However, it may be that a substantial fraction of capital at any time must be invested to achieve the "optimal" growth process. We will consider now the investor's net return on investment defined as net capital gain divided by total capital invested. Formally we have net return on investment to period T given by

$$R_T(\omega^T, \gamma^T) = (F_T(\omega^T, \gamma^T) - F_0)/\sum_{t=1}^{T} \sum_{i=1}^{n} \gamma_{ti} F_{t-1}(\omega^{t-1}, \gamma^{t-1}) \qquad (3.3)$$

Using equation (2.1) we can rewrite (3.3) as

$$R_T(\omega^T, \gamma^T) = \frac{1/F_0}{\sum_{t=1}^{T}(\sum \gamma_{T-t+1})(1/F_t(\omega^t, \gamma^{T-t+1}))} - \frac{F_0}{\sum_{t=1}^{T}\sum_{i=1}^{n}\gamma_{ti}F_{t-1}(\omega^{t-1}, \gamma^{t-1})} \qquad (3.4)$$

where $\gamma^{T-t+1} = (\gamma_{T-t+1}, \dots, \gamma_T)$, and $\sum \gamma_{T-t+1} = \sum_{i=1}^{n} \gamma_{T-t+1,i}$.

From this equation we get the following result.

Theorem 3.5 (Ethier and Tavaré)

If the strategy $\{\gamma_t\}$ is such that $F_T(\omega^T, \gamma^T)$ converges almost surely to ∞ with T, then $R_T(\omega^T, \gamma^T)$ converges in distribution to

$$R(\omega^\infty, \gamma^\infty) = 1/F_0 \ \lim_{T\to\infty} \sum_{t=1}^{T}(\sum \gamma_{T-t+1})(1/F_t(\omega^t, \gamma^{T-t+1})) \qquad (3.5)$$

Of course since ϕ is **favorable** there is at least one strategy such that F_t converges to ∞ a.s., and then from theorem 3.4 there is an infinite number of strategies for which this is true. We are interested in the return on investment for these **positive growth** strategies. A strategy has two components: (i) the total invested in each period and (ii) the distribution within a period of this total across the investment opportunities. Let the total investment at time t be given by $\Gamma_t = \sum_{i=1}^{n} \gamma_{ti}$ and the distribution at time t be $\pi_t = 1/\Gamma_t \cdot \gamma_t$. Since we can do as well in net return with a stationary strategy as with any other strategy we will restrict attention to that class.

Theorem 3.6 (Ethier and Tavaré)

Consider the stationary strategy $\gamma_t = \gamma(t)$ and the net return $R(\omega^\infty, \gamma)$. Then

(i) $\text{Prob}[0 \leq R \leq F_0] = 1$

(ii) $E R \leq \pi \cdot E\phi$.

The surprising fact in the above result is that the expected net return in the long run is less than the expected one period net return when we use a stationary strategy. To clarify this a bit consider the expression

$$\bar{R}_T(\gamma^T) = \frac{E(F_T(\omega^T, \gamma^T) - F_0)}{E \sum_{t=1}^{T} \sum_{i=1}^{n} \gamma_{ti} F_{t-1}(\omega^{t-1}, \gamma^{t-1})} \quad . \tag{3.6}$$

Theorem 3.7   (Griffin)

Suppose we have a stationary strategy $\gamma_t = \gamma(t)$.   **Then** $\bar{R}_T(\gamma) = \pi \cdot E\phi$.

So the expected one period net return is the same as the expected gain over the total expected investment.

4.   SECURITY

We have seen that the capital at a point in time $F_t(\gamma, \phi)$ depends upon a deterministic investment strategy γ and a stochastic return function ϕ . In our discusssion of growth in the previous section we concentrated on the mean of the distribution for $F_t(\gamma, \phi)$, or at least the mean of $Z_t(\gamma, \phi) = \log F_t(\gamma, \phi)$. We now turn attention to characteristics of the distribution of $Z_t(\gamma, \phi)$ which relate to security or risk aversion.

Before defining measures of security it is worth noting that the stochastic return function ϕ may depend upon a number of factors which can be assessed. Knowledge of these factors could for example decrease the variance of ϕ and thereby increase the security of strategies. For example, in blackjack information on the cards remaining in the deck (recorded by the Thorpe ten count) is known to improve the mean and variance of returns. In our analysis it is important that the return function is stationary (playing the same game each time), so we will assume that information on relevant factors remains constant over time, and has been incorporated into the definition of the return function ϕ .

Suppose, then, we have the wealth $F_t(\gamma^t, \omega^t)$ and consider $Z_t(\gamma^t, \omega^t) = \log F_t(\gamma^t, \omega^t)$. From (2.1) we get

$$Z_t(\gamma^t, \omega^t) = Z_{t-1}(\gamma^{t-1}, \omega^{t-1}) + J(\gamma_t, \omega_t) \qquad (4.1)$$

where $Z_t = \log F_0$, $J(\gamma_t, \omega_t) = \log(1 + \phi(\omega_t, \gamma_t))$, $t = 1, 2, \dots$. So Z_t, $t = 1, 2, \dots$ is a **random walk** with jump processes J .

In the previous section we considered policies $(\gamma_1, \gamma_2, \dots)$ which yielded a favorable random walk Z_t, $t = 1, 2, \dots$, in the mean. Consider, now, the following measures of security for Z_t, $t = 1, 2, \dots$.

$$\alpha(\gamma^\infty) = \text{Prob}\big[\tau_{\{Z_t(\gamma^t, \omega^t) \geq U\}} < \tau_{\{Z_t(\gamma^t, \omega^t) \leq L\}}\big] \quad , \qquad (4.2)$$

where $\tau_{\{Z_t(\gamma^t, \omega^t) \geq U\}}$ is the first passage time to the set $[U, \infty)$ in R .

$$\beta(\gamma^\infty) = \text{Prob}\big[Z_t(\gamma^t, \omega^t) \geq b_t, \ t = 1, 2, \dots\big] \quad , \qquad (4.3)$$

where b_t, $t = 1, 2, \dots$ is a specified lower bound on the log growth path. For any investment strategy $\gamma^\infty = (\gamma_1, \gamma_2, \dots)$ we can assess the security measures $\alpha(\gamma^\infty)$ and/or $\beta(\gamma^\infty)$ and consider mean growth and security in the selection of a strategy. We will illustrate this with the Bernoulli case $(\Omega = \{0,1\})$ and a single investment opportunity. So the random walk Z_t, $t = 1, 2, \dots$ has jump process given by

$$J(\gamma_t) = \begin{cases} \log(1 + \gamma_t \phi_0) & \text{w.p. } p_0 \\ \log(1 + \gamma_t \phi_1) & \text{w.p. } p_1 \end{cases} .$$

Theorem 4.1

Suppose we have a stationary strategy γ and consider the random walk $Z_t(\gamma)$, $t = 1, 2, \dots$, starting from $Z_0 \in R$. Then we have for $\theta > 1$ such that $p_0 \theta^{\log(1+\gamma\phi_0)} + p_1 \theta^{\log(1+\gamma\phi_1)} = 1$,

$$\frac{\theta^{Z_0} - \theta^L}{\theta^{U - \log(1+\gamma\phi_0)} - \theta^L} \leq \alpha(\gamma) \leq \frac{\theta^{Z_0} - \theta^{L - \log(1+\gamma\phi_1)}}{\theta^U - \theta^{L - \log(1+\gamma\phi_1)}} .$$

Proof: This is an adaptation of the theorem on p. 334 in Feller [3].

Theorem 4.2

With the stationary strategy γ and the random walk $Z_t(\gamma)$, $t = 1, 2, \dots$, starting from $Z_0 \in R$, suppose we have the lower bounds $b_t = -\infty$ $(t \neq T)$, b_T finite. Then with

$$W(\gamma) = \frac{b_T - Z_0 - T \log(1 + \gamma\phi_0)}{\log \left(\dfrac{1 + \gamma\phi_1}{1 + \gamma\phi_0}\right)}$$

we have

$$B(\gamma) \approx \Phi\left(\frac{W(\gamma) - T\,p_1}{\sqrt{T}\;p_0 p_1}\right) \quad ,$$

where Φ is the cumulative normal.

Proof: With the given bounds we get $B(\gamma) = \text{Prob}[z_T \geq b_T]$. Furthermore, with $\Omega = \{0,1\}$ we have $z_T = z_0 + W(\gamma)\log(1 + \gamma\phi_1) + (T - W(\gamma))\log(1 + \gamma\phi_0)$ where the number of 1's, $W(\gamma)$, is binomial. The result follows from the normal approximation to the binomial.

An illustration of theorem 4.1 for the game of blackjack is given in figure 4.1.

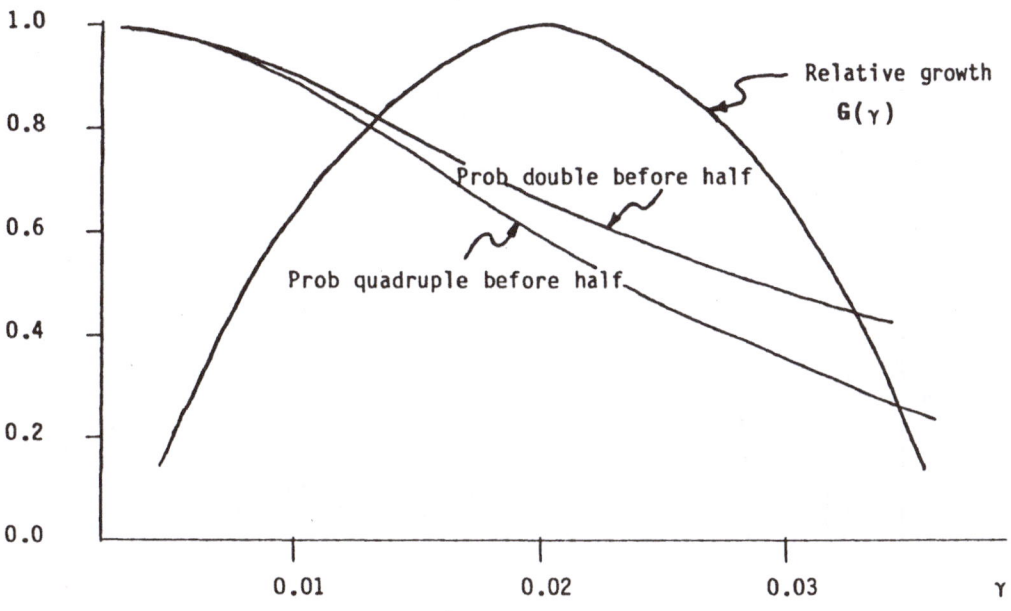

Figure 4.1: $\alpha(\gamma)$ and $G(\gamma)$ for blackjack example

5. GROWTH/SECURITY TRADE OFF

We will complete our discussion of the risky investment model by considering the trade off between growth and security, that is, between $G(\gamma)$ and $\alpha(\gamma)$ (or $B(\gamma)$).

In defining an index we assume that (i) $\alpha(\gamma)$ is monotone decreasing in γ and (ii) $G(\gamma)$ is concave in γ. So we can limit attention to strategies $\gamma \leq \bar{\gamma}$, $\bar{\gamma}$ = Kelly strategy. In the following definition ∇G and $\nabla \alpha$ refer to the Gateaux diferential.

$$I(\gamma|\gamma^*,\gamma^{**}) = \frac{\nabla G(\gamma)}{\nabla \alpha(\gamma)} \cdot \frac{\alpha(\gamma^*)}{G(\gamma^{**})} \quad , \qquad \gamma \leq \bar{\gamma} \quad , \tag{5.1}$$

where γ^* and γ^{**} are strategies chosen to standardize security and growth respectively. There are natural choices for γ^*, γ^{**}; for example (i) $\gamma^* = \gamma^{**} = \bar{\gamma}$, where growth and security are relative to Kelly growth and security; (ii) $\gamma^* = \varepsilon \to 0$, $\gamma^{**} = \bar{\gamma}$, where growth and security are relative to maximum growth and security.

To illustrate the trade off index we return to the blackjack example.

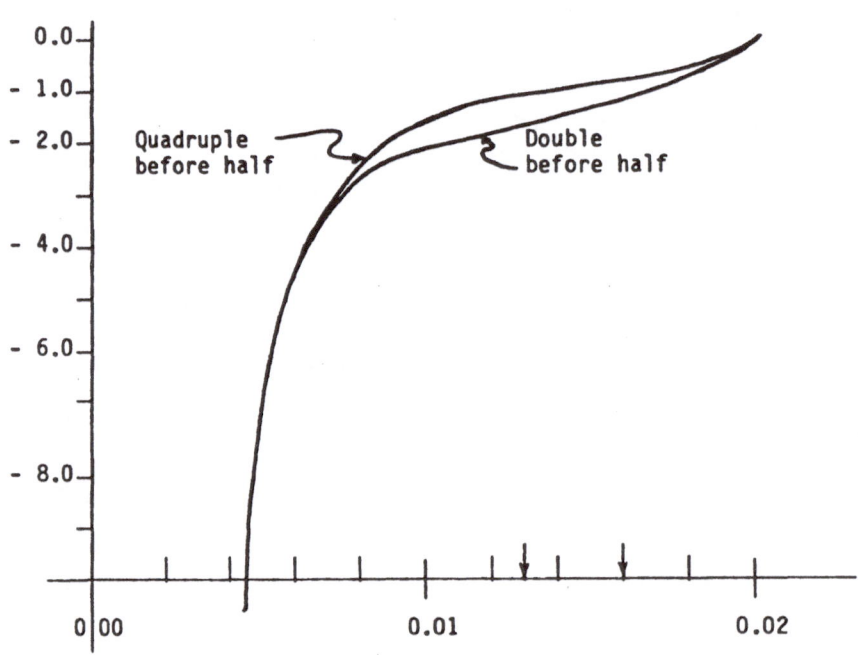

Figure 5.1: $I(\gamma|\varepsilon,\bar{\gamma})$ for blackjack example

If we look for the equilibrium value:

$$I(\gamma|\varepsilon,\bar{\gamma}) = -1$$

then the appropriate strategies become .016 and .013 for the double before half
and quadruple before half criteria respectively. These figures correspond to 80% and
65% of the Kelly strategy. If the criteria were stricter the percentage would
decrease further. It is worth noting that professional blackjack teams wager between
20% and 80% of the Kelly strategy.

REFERENCES

[1] Breiman, L. "Optimal gambling systems for fair games". Proc. 4th Berkeley
       Symp. on Math. Stat. and Prob., 1 (1961), 65-68.
[2] Ethier, S.N. and S. Tavaré. "The proportional bettor's return on investment".
       J. of Appl. Prov., 20 (1983), 563-573
[3] Feller, W. An Introduction to Probability Theory and its Applications.
       Vol. I, 2nd end. New York: John Wiley & Sons Inc., 1962.
[4] Finkelstein, M. and R. Whitley. "Optimal strategies for repeated games". Adv.
       Appl. Prob., Vol. 13 (1981), 415-428.
[5] Griffin, P. "Different measures of win rate for optimal proportional betting".
       To appear in Management Science (1983).
[6] Hausch, D., W. Ziemba and M. Rubinstein. "Efficiency of the market for race
     track betting". Management Science Vol. 27, No. 12 (1981), 1435-1452.
[7] Kelly, J. "A new interpretation of information rate". Bell System Tec. J.,
       Vol. 35 (1956), 917-926.

QUEUE PREDICTORS FOR STOCHASTIC TRAFFIC FLOWS CONTROL (++)

M.Peruggia (*)   F.Schoen (+)   M.G.Speranza (*)

1. Introduction

On-line control of semaphorized intersections is a major
step towards good performance of a traffic network. Efficiency
of on-line control is based upon general and syntetic modelling,
accurate prediction of queue evolution and good control strategies.

In Baras, Levine and Lin (1979) and Betrò, Schoen and Spe-
ranza(1983) stochastic dynamical models for traffic problems
were presented and formulas for prediction of queue length developed. Time was discretized so that no more than one vehicle
was assumed to cross a detecting line in a unit time interval.
This assumption makes it necessary to assume a thick discretization of time and gives rise to computational problems in the
actual control implementation; besides, it forces the model to
consider only single-lane streets. Moreover, the arrival rate
was assumed to be dependent only upon time and queue length and
a priori known.

In this paper the assumption on time discretization is relaxed
and any number of vehicles is allowed to be detected in a unit
time interval; so, multiple-lane streets can be considered. No

(*) Dipartimento di Matematica-Università di Milano
(+) Istituto per le Applicazioni della Matematica e dell'Informatica-C.N.R.-
    Milano
(++) This research has been partially supported by C.N.R. - Progetto Finaliz-
    zato Trasporti.

assumptions will be made upon dependences of arrival rate, so that, for example, it could be dependent also upon past arrival rates. Moreover, the arrival rate is not required to be known, but its on-line estimation is used to predict queue evolution.

2. Prediction in a partially observed process

Let us consider a discrete time process $<\lambda(t),z(t),n(t)>$, where $n(t)$ is observable and $\lambda(t)$ and $n(t)$ are non-observable. The components $\lambda(t)$, $z(t)$, $n(t)$ are stochastic processes whose state spaces, which we assume, for sake of simplicity, to be discrete, are respectively denoted by L, Z, N.

We are interested in finding expression for the predictor

$$P(z(t+1)=\bar{z}(t+1)\,|\,n^{t}=\bar{n}^{t}) \tag{2.1}$$

where $\bar{z}(t+1)\in Z$ and $n^{t}=(n(0),\ldots,n(t))$ and $\bar{n}^{t}=(\bar{n}(0),\ldots,\bar{n}(t))$ with $\bar{n}(i)\in Z$, $i=0,\ldots t$.

$$P(z(t+1)=\bar{z}(t+1)\,|\,n^{t}=\bar{n}^{t}) =$$

$$= \sum_{\bar{\lambda}(t+1)\in L} P(\lambda(t+1)=\bar{\lambda}(t+1),z(t+1=\bar{z}(t+1)\,|\,n^{t}=\bar{n}^{t})$$

$$= \sum_{\bar{\lambda}(t+1)\in L}\,\sum_{\bar{\lambda}(t)\in L}\,\sum_{\bar{z}(t)\in Z} P(\lambda(t+1)=\bar{\lambda}(t+1),z(t+1)=\bar{z}(t+1)\,|$$

$$\lambda(t)=\bar{\lambda}(t),z(t)=\bar{z}(t),n^{t}=\bar{n}^{t})\,P(\lambda(t)=\bar{\lambda}(t),z(t)=\bar{z}(t)\,|\,n^{t}=\bar{n}^{t})$$

$$\tag{2.2}$$

Expression (2.2) contains two probabilities, the first of which is assumed to be known; the second one can be given a recursive expression.

$$P(\lambda(t)=\bar{\lambda}(t),z(t)=\bar{z}(t)\,|\,n^{t}=\bar{n}^{t}) =$$

$$= \frac{P(\lambda(t)=\bar{\lambda}(t),z(t)=\bar{z}(t),n(t)=\bar{n}(t)\,|\,n^{t-1}=\bar{n}^{t-1})}{P(n(t)=\bar{n}(t)\,|\,n^{t-1}=\bar{n}^{t-1})}$$

$$= \frac{P(\lambda(t)=\bar{\lambda}(t),z(t)=\bar{z}(t),n(t)=\bar{n}(t)\mid n^{t-1}=\bar{n}^{t-1})}{\displaystyle\sum_{\bar{\lambda}(t)\in L}\ \sum_{\bar{z}(t)\in Z}\ P(\lambda(t)=\bar{\lambda}(t),z(t)=\bar{z}(t),n(t)=\bar{n}(t)\mid n^{t-1}=\bar{n}^{t-1})} .$$

$$P(\lambda(t)=\bar{\lambda}(t),z(t)=\bar{z}(t),n(t)=\bar{n}(t)\mid n^{t-1}=\bar{n}^{t-1}) =$$

$$= \sum_{\bar{\lambda}(t-1)\in L}\ \sum_{\bar{z}(t-1)\in Z}\ P(\lambda(t)=\bar{\lambda}(t),z(t)=\bar{z}(t),n(t)=\bar{n}(t)\mid \lambda(t-1)=\bar{\lambda}(t-1),$$

$$z(t-1)=\bar{z}(t-1),n^{t-1}=\bar{n}^{t-1})\,P(\lambda(t-1)=\bar{\lambda}(t-1),z(t-1)=\bar{z}(t-1)\mid n^{t-1}=\bar{n}^{t-1})$$

$$= \sum_{\bar{\lambda}(t-1)\in L}\ \sum_{\bar{z}(t-1)\in Z}\ P(n(t)=\bar{n}(t)\mid \lambda(t)=\bar{\lambda}(t),\lambda(t-1)=\bar{\lambda}(t-1),z(t)=\bar{z}(t),$$

$$z(t-1)=\bar{z}(t-1),n^{t-1}=\bar{n}^{t-1})\,P(\lambda(t-1)=\bar{\lambda}(t-1),z(t-1)=\bar{z}(t-1)\mid n^{t-1}=\bar{n}^{t-1})$$

$$P(\lambda(t)=\bar{\lambda}(t),z(t)=\bar{z}(t)\mid \lambda(t-1)=\bar{\lambda}(t-1),z(t-1)=\bar{z}(t-1),n^{t-1}=\bar{n}^{t-1})$$

$$(2.3)$$

Assuming that the first probability of expression (2.3) and the initial probability $P(\lambda(0)=\bar{\lambda}(0),z(0)=\bar{z}(0))$ are known, the recursion can be set up.

3. Prediction of queue evolution.

Let us consider a simple traffic model in which a traffic light controls a crossing of one-way streets and a detector counting vehicle passages is placed upstream the stop line on each street entering the intersection. Thus, for each street the observable component $n(t)$ of the general process of chapter 2 represents the number of vehicles detected during the unit time interval $t,t+1$; $\lambda(t)$ and $z(t)$ are, respectively, the arrival rate at the detector and the queue length at time t. The predictor (2.1) thus gives the probability distribution of the queue length, which is only partially observable through the arrival process $n(t)$.

Rewriting (2.2) in a more suitable way, we obtain:

$$P(\lambda(t+1)=\bar{\lambda}(t+1),z(t+1)=\bar{z}(t+1)\mid \lambda(t)=\bar{\lambda}(t),z(t)=\bar{z}(t),n^{t}=\bar{n}^{t})$$

$$= P(z(t+1)=\bar{z}(t+1)\mid \lambda(t+1)=\bar{\lambda}(t+1),\lambda(t)=\bar{\lambda}(t),z(t)=\bar{z}(t),n^{t}=\bar{n}^{t})$$

$$\cdot\ P(\lambda(t+1)=\bar{\lambda}(t+1)\mid \lambda(t)=\bar{\lambda}(t),z(t)=\bar{z}(t),n^{t}=\bar{n}^{t}) \qquad (3.1)$$

It is sensible to suppose that the quantities on the right-hand side in (3.1) are known to the modeller; the first quantity represents the probability of a transition from queue length $\bar{z}(t)$ to queue length $\bar{z}(t+1)$, while the second one represents the evolution of the arrival rate. We note that the transition probability for $z(t+1)$ should be based upon a suitable model for the (unobserved) departures from the stop line. It is also quite reasonable to suppose that the quantity

$$P(n(t)=\bar{n}(t)\mid \lambda(t)=\bar{\lambda}(t),\lambda(t-1)=\bar{\lambda}(t-1),z(t)=\bar{z}(t),z(t-1)=\bar{z}(t-1),$$
$$n^{t-1}=\bar{n}^{t-1}) \tag{3.2}$$

is known as well. For example, the arrival process can be modelled through a doubly stochastic Poisson process (see for example Bremaud (1981)); assuming that $n(t)$ in (3.2) depends only on $\lambda(t)$ we obtain:

$$P(n(t)=\bar{n}(t)\mid \lambda(t)=\bar{\lambda}(t),\lambda(t-1)=\bar{\lambda}(t-1),z(t)=\bar{z}(t),z(t-1)=\bar{z}(t-1),$$
$$n^{t-1}=\bar{n}^{t-1}) = P(n(t)=\bar{n}(t)\mid \lambda(t)=\bar{\lambda}(t))$$

$$= (\lambda(t))^{\bar{n}(t)}\exp(-\bar{\lambda}(t))/(\bar{n}(t))!$$

Let us suppose now that the following assumptions hold for the general model presented in this section:
1) the arrival rate process is known a priori (i.e., need not be estimated)
2) the time discretozation is such that no more than one vehicle can arrive at the detector or leave the intersection during the same unit time interval.

In this case equations (2.1), (2.2) and (2.3) reduce to the recursions given in Baras, Levine and Lin (1979) and in Betrò, Schoen and Speranza (1983). In fact, let us suppose first that $\bar{n}(t)=1$; than we obtain

$$P(z(t)=j \mid n(t)=1, n^{t-1}=\bar{n}^{t-1}) = \frac{P(z(t)=j, n(t)=1 \mid n^{t-1}=\bar{n}^{t-1})}{\displaystyle\sum_{i=0}^{N} P(z(t)=i, n(t)=1 \mid n^{t-1}=\bar{n}^{t-1})}$$

$$= \frac{P(n(t)=1 \mid z(t)=j, n^{t-1}=\bar{n}^{t-1}) \cdot P(z(t)=j \mid n^{t-1}=\bar{n}^{t-1})}{\displaystyle\sum_{i=0}^{N} P(n(t)=1 \mid z(t)=i, n^{t-1}=\bar{n}^{t-1}) \cdot P(z(t)=i \mid n^{t-1}=\bar{n}^{t-1})}$$

$$= \frac{\lambda(j,t)\, P(z(t)=j \mid n^{t-1}=\bar{n}^{t-1})}{\displaystyle\sum_{i=0}^{N} \lambda(i,t)\, P(z(t)=i \mid n^{t-1}=\bar{n}^{t-1})}$$

where $\lambda(i,t)=P(n(t)=1 \mid z(t)=i, n^{t-1}=\bar{n}^{t-1})$, $i=0,1,\ldots,N$, and N represents the maximum queue length.

If $\bar{n}(t)=0$, we have analogously:

$$P(z(t)=j \mid n(t)=0, n^{t-1}=\bar{n}^{t-1}) = \frac{(1-\lambda(j,t))\, P(z(t)=j \mid n^{t-1}=\bar{n}^{t-1})}{\displaystyle\sum_{i=0}^{N} (1-\lambda(i,t))\, P(z(t)=i \mid n^{t-1}=\bar{n}^{t-1})}$$

Moreover, we have:

$$P(z(t)=j \mid n^{t-1}=\bar{n}^{t-1}) = \sum_{i=0}^{N} P(z(t)=j \mid z(t-1)=i, n^{t-1}=\bar{n}^{t-1}) P(z(t-1)=i \mid n^{t-1}=\bar{n}^{t-1})$$

$$= \sum_{i=0}^{N} q_{ij} P(z(t-1)=i \mid n^{t-1}=\bar{n}^{t-1})$$

where

$q_{ii} = (1-\lambda(i,t))(1-\mu(i,t)) + \lambda(i,t)\mu(i,t)$

$q_{ii-1} = \lambda(i-1,t)(1-\mu(i-1,t))$

$q_{ii+1} = \mu(i+1,t)(1-\lambda(i+1,t))$

$q_{ij} = 0$ otherwise

$\mu(i,t) = P(\text{a departure at time } t \mid z(t)=i, n^{t-1}=\bar{n}^{t-1})$

Conclusions

Formulas for prediction in a partially observed stochastic process have been derived and given in a recursive form. It has been shown how these formulas can be applied tp the prediction of queue evolution in a traffic environment. Accurate and easily implementable predictors are basic tools for setting up an efficient control strategy of traffic flows regulated by a traffic light. Different control strategies based upon queue length predictors can be developed; the behaviour of some of them has been compared by means of simulation in Betro, Schoen and Speranza (1984).

The introduction, in this paper, of bulk arrivals allows the controller to assume for the optimization larger unit time intervals with great advantage for on-line implementation. Besides, the introduction of estimated (rather than a priori known) arrival rates is a first step towards modelling and control of simple traffic networks.

Acknowledgements

We are greatly indebted to Bruno Betrò for many valuable comments and suggestions.

References

Baras J.S., Levine W.S. and Lin T.L. (1979), Discrete time point processes in urban traffic queue estimation, I.E.E.E. Trans. Automat. Contr., AC-24, 12-27.

Betrò B., Schoen F. and Speranza M.G. (1983), Modelling and optimization of stochastic traffic flows, IAMI T.R. 83.8.

Betrò B., Schoen F. and Speranza M.G., Stochastic on-line control strategies for a semaphorized intersection, to appear in Proceedings of the 2nd National Meeting of Progetto Finalizzato Trasporti, Bologna.

Bremaud P. (1981), Point processes and queues - martingale dynamics, Springer-Verlag, New York.

ITERATIVE APPROXIMATIONS FOR NETWORKS OF QUEUES

Jan van Doremalen and Jaap Wessels

Eindhoven, 1983

Abstract. If networks of queues satisfy certain conditions, then the equilibrium
distribution for the number of jobs in the various stations has the so-called product-form. In such cases there are relatively elegant and simple computational procedures for the relevant behavioral characteristics. Quite commonly, however, the
conditions are too severe and exact solution is practically impossible for larger
problems.
In this paper we will consider iterative approximations for networks of queues which
either don't possess product-form solutions or are so large that exact solution becomes intractable even using the product-form of the solution. The approximations
are based on a mean value analysis approach and use either aggregation of some sort
or decomposition. For the details of the approximations heuristic arguments are used.
The approach is worked out for some problem types.

1. Introduction

In many areas networks of queues are used as models: production planning in manufacturing enterprises, computer performance evaluation, design of communication networks, planning of harbour facilities, etc. General queueing theory does not provide
much help for the analysis of such complex queueing models. The only help can be
found in the line of research that emerged from Jackson's paper [10] in which it
was proved that the equilibrium distribution for a particular type of networks has
a product-form. Extending Jackson's result it has been proved that a large class of
networks has equilibrium probabilities with a product-form (confer Kelly [11]). It
has also been shown that for such queueing networks the relevant behavioral characteristics can be computed in some (relatively) simple and elegant ways. The two main
procedures are known by the name of convolution method (confer Reiser and Kobayashi
[16]) and by the name of mean value analysis (confer Reiser and Lavenberg [17] and
Reiser [15]).

Regrettably, however, many practical problems do not satisfy the conditions for having
product-form solutions, whereas other problems are very large and therefore intractable using the standard methods. For both types of problems the only way out seems
to be approximation. Several methods of approximation have been published. For instance approximate decomposition (confer Courtois [4]) which is used in the handling
of memory queues in computer evaluation studies (confer Hine, Mitrani and Tsur [9])

and for handling FIFO-servers with arbitrary service time distributions in open networks (confer Kühn [12]). For an overview of several approaches see Chandy and Sauer
[3].

In recent years the mean value analysis procedure has become popular as a basis for
approximation. For a recent overview and appraisal see de Souza a Silva, Lavenberg
and Muntz [19]. Although the approximation methods for different types of problems
show some structural resemblance, the methods are basically heuristic. Only in some
cases one has been successful in obtaining convergence and uniqueness results (see
[19] for some examples and further references).
In this paper we will present heuristics and numerical results for two types of problems and discuss the same topics for some other problems.
The first problem, which will be treated in Section 2, is a rather specific one. It
arised in treating the planning of harbour facilities, where it appeared to be necessary to include servers with a two-phase service procedure. The first phase is a
preparatory one and may be executed for the first customer of a busy period in the
preceding idle period. This feature destroys the product-form.
The next problem has attracted a lot of attention in the literature: the problem of
many customer chains in a closed network. The conditions for the existence of a product-form solution are not violated, but even the efficient mean value analysis procedure requires too much work if the number of chains is relatively large. To obtain
approximations, the usual approach is to remove the recursion from the mean value
scheme. In Section 3, we will present a decomposition approach, which maintains the
recursion, but transforms the multidimensional recursion in several one-dimensional
recursions.
For both problems numerical results are compared to exact solutions. For the second
problem a comparison with other methods will be given also.
In Section 4 some experience with other methods will be reported. Here, as well as
in Section 2, the heuristics are basically some sort of aggregation. Disaggregation
provides the basis for the next iteration step.

2. The two-phase server with preparatory first phase.

Consider a closed queueing network with N single server FIFO stations in which K
customers walk around with routing probabilities p_{mn} for jumping from station m to
station n. At station n the customers have exponentially distributed workloads with
mean w_n. The network satisfies the conditions for having a product-form solution.
For such networks there is an arrival theorem stating that a customer sees upon a
junp or arrival moment the system as if in equilibrium with $K - 1$ customers. Using
this theorem we may evaluate steady-state quantities by setting up a recursive scheme.
The mean residence time $S_n(K)$ at queue n may be expressed in terms of the mean number

of customers at that queue, if there are K - 1 customers system, $L_n(K-1)$,

(1) $S_n(K) = L_n(K-1)w_n + w_n$.

The RHS denotes the average amount of work a customer sees in front of him upon arrival at queue n plus his own work. Applying Little's formula to queue n, we obtain with $\Lambda_n(K)$ being the throughput at queue n,

(2) $L_n(K) = \Lambda_n(K) S_n(K)$.

The throughput at queue n is the quotiënt of the number of customers in the system and the mean time for a round trip of a customer starting at queue n,

(3) $\Lambda_n(K) = K \left(\sum_{m=1}^{N} \frac{\vartheta_m}{\vartheta_n} S_m(K) \right)^{-1}$

where the visiting-ratio's ϑ_m are the unique solution of

(4) $\vartheta_m = \sum_{i=1}^{N} \vartheta_i \, p_{im}$ and $\sum_{m=1}^{N} \vartheta_m = 1$.

Note that ϑ_m/ϑ_n denotes the mean number of visits to queue m per visit to queue n.

Starting with $L_n(0) = 0$ these relations give a recursive scheme to evaluate the mean values. For more details on this mean value scheme and the arrival theorem we refer to Reiser and Lavenberg [17] and Reiser [15]. If we introduce an extraordinary behaviour at one of the stations, for example non-exponential service times, formulae (2), (3) and (4) remain valid. However, relation (1) will be violated. To some extent the idea behind the relation will remain and, therefore, it seems sensible to consider a mean value scheme with a slightly adjusted form of relation (1) to incorporate the effects of the extraordinary behaviour.

As an example of such a deviant behaviour we will consider a network where some server n may have a workload, which per customer consists of two negative exponentially distributed phases, $w_n = w'_n + w''_n$. The first phase is a kind of preparatory one and can be started (and sometimes be completed) during an idle period. Thus the first customer of a busy period has a different workload and the effect will be that some of the customers only experience a workload w''_n, whereas others have the full workload $w'_n + w''_n$.

The steady-state probabilities no longer have a product-form, but the network still can be analyzed as a continuous-time Markov-process on a finite state space. To solve for the corresponding set of equilibrium equations is very unattractive from a computational point of view. We will develop an iterative approximation based on the mean value scheme and an adjustment of relation (1).

The first guess in adapting Formula (1) seems to be to maintain $L_n(K-1)$ as the expected number of customers present upon arrival (this need not be true) and to replace w_n by an adjusted value,

$$(5) \qquad \tilde{w}_n = (1 - a_n)w'_n + w''_n ,$$

where a_n denotes the probability that an arriving customer finds his preparatory phase already completed. Thus we implicitly assume that all customers have the same negative exponentially distributed workload with mean $\tilde{w}_n$, i.e. we approximate the original model by a model with a product-form solution. To find a_n requires a rigorous analysis of the original problem and that we just wanted to avoid. However, one may make a guess, for instance $a_n = 0$ or $a_n = 1$, and try to improve the guess after an evaluation of the mean value scheme. Suppose we have an initial guess for a_n and we have solved the mean value scheme (1) through (4) with w_n replaced by $\tilde{w}_n$. How to improve on the initial guess for a_n? The true a_n can be written as

$$(6) \qquad a_n = b_n c_n$$

with b_n the probability that an arriving customer is the first one in a busy period and c_n the probability that a preparatory phase is completed before the end of an idle period. Better estimates for b_n and c_n then can be constructed as follows (confer van Doremalen and Wessels [7]),

$$(7) \qquad b'_n = 1 - \Lambda_n(K-1)\tilde{w}_n$$

$$(8) \qquad c'_n = w'_n (w'_n + v'_n)^{-1}$$

where

$$(9) \qquad v'_n = (1 - \Lambda_n(K)\tilde{w}_n)(\Lambda_n(K)b'_n)^{-1} .$$

The results of the iteration scheme are fairly good, particularly for the throughput. Mean queue lengths and mean residence times in general are less accurately approximated. As a simple numerical example the results of a cyclical network with three stations are depicted in Table 1. Evaluated are the exact throughput, a lowerbound $(a_1 = 1)$, an upperbound $(a_1 = 0)$ and the approximation resulting from the iterative method. The last column gives the limiting values for a_1.

w_2 w_3	throughputs				a_1
	exact	low	appr.	high	
2 2	.326	.300	.321	.347	.42
8 8	.093	.091	.092	.093	.83
.25 .25	.500	.497	.499	.946	.01

Table 1. Throughputs in a cyclical network with one two-phase server $w_1' = w_1'' = 1$, $K = 3$ and $N = 3$.

It is possible to refine these results. One way would be to use Kühns decomposition approach, confer Kühn [12], to take into account the non-exponential character of the two phase servers. A natural extension of the method then is to consider the case that the phases themselves are non-exponential.

3. Closed multichain queueing networks

Again consider a closed network with N single server FIFO stations. Now there are R irreducible customer chains, where the K_r customers of chain r have routing proba-bilities p_{mn}^r for going from station m to station n. At station n all customers have negative exponentially distributed workloads with the same expected value w_n. The arrival theorem states that a customer sees upon a jumpmoment the system in equili-brium as if one customer of his own chain has been removed. If we denote the popu-lation-vector $(K_1,\ldots,K_r)$ as K, this theorem implies that $S_{nr}(K)$, the mean residence time of a chain r customer at station n, can be expressed in $L_{n\ell}(K - e_r)$, the mean number of chain ℓ customers at station n if one customer of chain r has been removed from the system,

$$(10) \qquad S_{nr}(K) = \sum_{\ell=1}^{R} L_{n\ell}(K - e_r)w_n + w_n .$$

Application of Little's formula to station n gives,

$$(11) \qquad L_{nr}(K) = \Lambda_{nr}(K) S_{nr}(K) ,$$

where $\Lambda_{nr}(K)$ denotes the throughput of chain r customers at queue n. Finally, the multichain-equivalent of Relation (3) is,

$$(12) \qquad \Lambda_{nr}(K) = \vartheta_{nr} K_r \left(\sum_{m=1}^{N} \vartheta_{mr} S_{mr}(K) \right)^{-1}$$

where the ϑ_{nr}'s are, for $r = 1,2,\ldots,R$, the unique solution of

$$(13) \qquad \vartheta_{nr} = \sum_{m=1}^{N} \vartheta_{mr} P_{mn}^{r} \quad \text{and} \quad \sum_{n=1}^{N} \vartheta_{nr} = 1 .$$

For more details on the multichain mean value scheme we refer to Reiser and Laven-berg [17].

The recursion, defined by the Relations (10) through (13), now runs through all vec-tors in the range from $(0,\ldots,0)$ to $(K_1,\ldots,K_R)$. The storage requirements and the complexity of the algorithm grow exponentially with the number of chains. The appa-rent problem differs essentially from the one described in Section 2. Now the pro-duct-form solution is not violated, but the complexity of the algorithm prohibits an exact evaluation for larger values of $R,K_1,\ldots,K_R$ and approximate methods have to be recommended for that reason.

In the literature several approximation methods have been considered, e.g. by Schweitzer [18], Reiser [14], Reiser and Lavenberg [17] and Chandy and Neuse [2]. Very recently, an overview of these and other methods appeared in de Souza a Silva, Lavenberg and Muntz [19]. The usual approach is to remove the recursion from the mean value scheme and to concentrate on an iterative approximation of the mean values at the population vector K. We will exploit a decomposition idea in which R single chain networks are analyzed. Iteratively an improved approximation of the mutual influence of the chains is incorporated in the single chain analysis.

For chain r, $r = 1,2,\ldots,R$, consider the following adjusted single chain mean value scheme. Evaluate for $k = 1,2,\ldots,K_r$,

$$(14) \qquad S_{nr}^{*}(k) = L_{nr}^{*}(k-1)w_n + w_n + A_{nr}(k)w_n$$

$$(15) \qquad \Lambda_{nr}^{*}(k) = \vartheta_{nr} k \left(\sum_{m=1}^{N} \vartheta_{mr} S_{mr}^{*}(k) \right)^{-1}$$

$$(16) \qquad L_{nr}^{*}(k) = \Lambda_{nr}^{*}(k) S_{nr}^{*}(k)$$

where the factor $A_{nr}(k)$ denotes the number of customers of other chains a chain r customer sees in front of him upon arrival at station n if k customers of his own chain are in the system. As an approximation for $A_{nr}(k)$ we propose

(17) $$A_{nr}(k) = \sum_{\ell \neq r} L^*_{n\ell}(K_\ell)$$

where we use as an approximation assumption that a chain r customer sees the other chains as if in global equilibrium. Equations (14) through (17) implicitly give the approximations for the mean values. A standard technique to solve for these equations is to start with initial values for the $A_{nr}(k)$'s and to iterate the scheme for the successive chains until convergence is established. Using Brouwer's Fixed Point Theorem one can prove the existence of a positive solution of the equations. Up till now we have not been able to prove uniqueness of the solution and convergence of the method. However, numerical experiments show a relatively fast convergence and the approximations usually are within a few percent of the exact values. One can construct examples where the approximations are rather poor.

We will show a numerical example where we have compared the exact results of the mean value scheme with four different approximation methods.

Consider the model of a computer system with three terminal groups pictured in Figure 1. The system consists of a central processor unit (CPU) and three disk-groups (D1, D2 and D3). The service discipline at these four stations is first-in first-out and the exponential workloads have expected values 10 m sec, 20 m sec, 20 m sec and 30 m sec respectively. There are three terminal groups (T1, T2 and T3). The 20 active terminals of T1 have mean think times of 10 sec. They generate requests which in the average have 20 CPU calls, 15 D1 calls and 4 D2 calls. A terminal starts thinking again if his request has been handled and a response has been returned. The 10 active terminals of T2 have thinktimes of 20 sec, and requests of 40 CPU calls, 14 D1 calls abd 25 D2 calls. The 10 active terminals of T3 have thinktimes of 60 sec and requests of 200 CPU calls, 20 D1 calls, 40 D2 calls and 139 D3 calls.

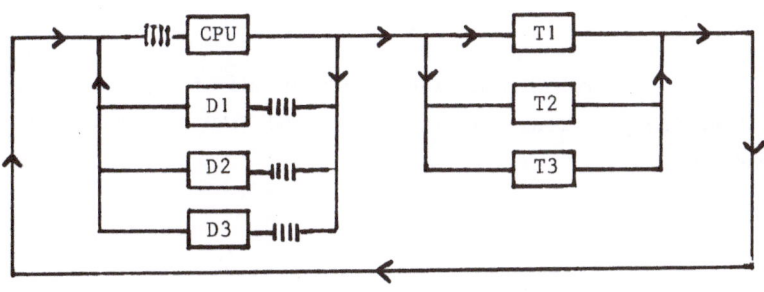

Figure 1. A computer system with terminal groups.

In Table 2 are pictured the utilizations of the CPU and the disk groups. The exact results are compared with four approximation methods, namely the methods of Schweitzer (SCHW), Reiser (R), Reiser and Lavenberg (R+L) and our method (D+W). In Table 3 are the response times for the three terminal groups. We note that the four methods all perform very good. At the moment we are studying other applications and examples and a more detailed report on the methods is in preparation. We remark that our method very straightforwardly can be extended to systems with LIFO (last-in first-out), PS (processor-sharing) and IS (infinite server) stations. This, for example, makes it possible to consider the above model with a processor-sharing CPU and consequently with different workloads at the CPU for the different terminalgroups. Finally, it should be noted that the method can be extended to mixed open and closed networks, confer van Doremalen [6].

	exact	SCHW	R	R + L	D + W
CPU	.774	.766	.770	.774	.768
D1	.686	.679	.683	.687	.682
D2	.457	.452	.454	.457	.454
D3	.539	.532	.536	.537	.535

Table 2. Utilizations in the computer system.

	exact	SCHW	R	R + L	D + W
T1	1.69	1.81	1.74	1.67	1.77
T2	3.11	3.29	3.22	3.02	3.26
T3	17.30	18.44	17.79	17.67	17.95

Table 3. Response times of the terminal groups.

4. Some other applications

In this section we will venture on some other applications of approximation techniques involving blocking phenomena, priority rules and FIFO stations with class dependent workloads.

4.1. Blocking

Consider the network model of Section 2. Now at queue n only a restricted number b_n of customers is allowed for. The joining of queue n is forbidden as long as b_n customers are present. A customer not allowed to enter station n, waits in the originating server and blocks this server until the unblocking moment. The effect of blocking is a decrease in the availability of the blocked servers. This can be accounted for by increasing the workloads at the blocked servers with some factor which may be determined iteratively, using estimates for the blocking probability from the preceding analysis. The results obtained so far, show an improvement compared to the total neglection of blocking effects. Especially, if the effect of blocking is not too heavy the approach seems to work quite well. A detailed report on this case is in preparation.

4.2. Priorities

Consider the model of Section 3. However, now there is some kind of priority for certain chains at certain queues. We thereby can think of preemptive-resume priorities and head-of-the-line priorities. Non-iterative approximations for such models for example can be based on the mean value analysis of M/G/1-priority queues as described in van Doremalen [5]. Iterative approximations might be based on the convolution algorithm. Results obtained so far are very promising and research in that direction is in progress.

4.3. Chain dependent workloads at FIFO single server stations

The mean value scheme of Section 3 for closed multichain queueing networks works alright if we assume the same negative exponentially distributed service times for all customer chains at a specific station. However, if the mean service times w_{nr} for the chains at a certain queue n do differ, the product-form solution no longer holds and the mean value scheme does not give exact results.

One way out is a relaxation of the mean value scheme. This straightforward non-iterative approximation has been considered by others also, confer Bard [1]. Instead of Relation (10) we get

$$(18) \qquad S_{nr}(K) = \sum_{\ell=1}^{R} L_{n\ell}(K - e_r)w_{n\ell} + w_{nr} \, .$$

Another method is the well-known processor-sharing approximation which reduces to
the following, intuitively less attracting, adjustment of (10),

$$(19) \qquad S_{nr}(K) = \sum_{\ell=1}^{R} L_{n\ell}(K - e_r)w_{nr} + w_{nr} \; .$$

Numerical experiments show the first method to be considerably better. A totally
different approach is to use a negative exponential service time distribution with
a mean which is a proper mixture of the original means. Iteratively, this mixture
can be determined. The results are not too well and it seems better to use explicit
estimates for the probability that the server works on a particular type of job.
A report on such an approach is in preparation.

5. Concluding remarks

We have considered the use of iterative approximation methods in several applications
The importance of approximation methods in the analysis of queueing networks is
paramount for several reasons.

First of all, exact analysis is limited to only a few restricted models as for example
the networks which satisfy the conditions for the existence of a product-form solution
for the steady-state probabilities. Though this class of networks still is subject of
research and techniques are being developed to extend the class (confer Kelly [11]
and van Dijk and Hordijk [8]), it is clear that very important classes of networks
never will be fitted in this frame.

But, as we have seen in Section 3, there is another problem. Even for models in a
class which can be analyzed elegantly, the amount of work to be done can prohibit
an exact evaluation of important performance measures. Of course, one can try to
improve the evaluation methods as for instance has been done by Lam and Lien [13],
but again there always will be the need of fast approximation methods.

References

[1]   Y. Bard, Some extensions to multichain queueing network analysis.
           4th Int. Symp. on Modelling and Performance Evaluation of Computer Sys-
           tems, Vienna 1979.

[2]   K.M. Chandy and D. Neuse, Lineariser: A heuristic algorithm for queueing net-
           work models of computing systems.
           Comm. of the A.C.M. 25 (1982) 126 - 134.

[3]   K.M. Chandy and C.H. Sauer, Approximate methods for analyzing queueing net-
        work models of computing systems.
        Computing Surveys 10 (1978) 281 - 317.

[4]   P.J. Courtois, Decomposability: Queueing and Computer System Applications.
        Academic Press, New York 1977.

[5]   J. van Doremalen, A mean value approach for M/G/1 priority queues.
        Memorandum COSOR 83-09, Eindhoven University of Technology 1983.

[6]   J. van Doremalen, Mean value analysis in multichain queueing networks: an
        iterative approximation.
        DGOR Operations Research Proceedings 1983, Springer Verlag, Berlin.
        To appear.

[7]   J. van Doremalen and J. Wessels, An iterative approximation for closed
        queueing networks with two-phase servers.
        Memorandum COSOR 83-12, Eindhoven University of Technology 1983.

[8]   N. van Dijk and A. Hordijk, Networks of queues: Part I, Job-local-balance
        and the adjoint process. Part II, General routing and service characte-
        ristics.
        Proc. of the Int. Sem. on Modelling and Performance Evaluation Methodo-
        logy, Paris 1982.

[9]   J.H. Hine, I. Mitrani and S. Tsur, The control of response times in multi-
        class systems by memory allocation.
        Comm. of the A.C.M. 22 (1979) 415 - 424.

[10] J.R. Jackson, Networks of waiting lines.
        O.R. 5 (1957) 518 - 521.

[11] F.P. Kelly, Reversibility and stochastic networks.
        John Wiley and Sons, New York 1978.

[12] P.J. Kühn, Approximate analysis of general queueing networks by decomposition.
        IEEE Trans. Comm. 27 (1979) 113 - 126.

[13] S. Lam and Y. Lien, A tree convolution algorithm for the solution of queueing
        networks.
        Comm. of the A.C.M. 26 (1983) 203 - 215.

[14] M. Reiser, Mean value analysis: A new look at an old problem.
        4th Int. Symp. on Modelling and Performance Evaluation of Computer Sys-
        tems, Vienna 1979.

[15] M. Reiser, Mean value analysis and convolution method for queue-dependent
        servers in closed queueing networks.
        Performance Evaluation 1 (1981) 7 - 18.

[16] M. Reiser and H. Kobayashi, Queueing networks with multiple closed chains:
        theory and computational algorithms.
        IBM J. Res. Dev. 19 (1975) 283 - 294.

[17] M. Reiser ans S.S. Lavenberg, Mean value analysis of closed multichain
        queueing networks.
        Comm. of the A.C.M. 27 (1980) 313 - 322.

[18] P. Schweitzer, Approximate analysis of multiclass networks of queues.
      Presented at the Int. Conf. on Stochastic Control and Optimization,
      Amsterdam 1979.

[19] E. de Souza a Silva, S.S. Lavenberg and R.R. Muntz, A perspective on itera-
      tive methods for the approximate analysis of closed queueing networks.
      Proc. Int. Workshop on Applied Mathematics and Performance Reliability
      Models of Computer Communication Systems, University of Pisa 1983.

CONVERGENCE THEORIES OF DISTRIBUTED ITERATIVE PROCESSES: A SURVEY[†]

by

Dimitri P. Bertsekas*

John N. Tsitsiklis**

Michael Athans*

Abstract

We consider a model of distributed iterative algorithms whereby several processors participate in the computation while collecting, possibly stochastic information from the environment or other processors via communication links. Several applications in distributed optimization, parameter estimation, and communication networks are described. Issues of asymptotic convergence and agreement are explored under very weak assumptions on the ordering of computations and the timing of information reception. Progress towards constructing a broadly applicable theory is surveyed.

[†] The research of D.P. Bertsekas was supported by NSF-ECS-8217668 and under DARPA Grant ONR-N00014-75-C-1183. The research of J.N Tsitsiklis and M. Athans was supported by ONR-N00014-77-C-0532(NR- 041-519).

*Dept. of Electrical Engineering and Computer Science, Laboratory for Information and Decision Systems, M.I.T., Cambridge, Mass. 02139.

**Dept. of Electrical Engineering, Stanford University, Stanford, California.

1. Introduction

Classical (centralized) theories of decision making and computation deal with the situation in which a single decision maker (man or machine) possesses (or collects) all available information related to a certain system and has to perform some computations and/or make a decision so as to achieve a certain objective. In mathematical terms, the decision problem is usually expressed as a problem of choosing a decision function that transforms elements of the information space into elements of the decision space so as to minimize a cost function. From the point of view of the theory of computation, we are faced with the problem of designing a serial algorithm which actually computes the desired decision.

Many real world systems however, such as power systems, communication networks, large manufacturing systems, public or business organizations, are too large for the classical model of decision making to be applicable. There may be a multitude of decision makers (or processors), none of which possesses all relevant knowledge because this is impractical, inconvenient, or expensive due to limitations of the systems's communication channels, memory, or computation and information processing capabilities.

In other cases the designer may deliberately introduce multiple processors into a system in view of the potential significant advantages offered by distributed computation. For problems where processing speed is a major bottleneck distributed computing systems may offer increases in throughput that are either unattainable or prohibitively expensive using a single processor. For problems where reliability or survivability is a major concern distributed systems can offer increased fault tolerance or more graceful performance degradation in the face of various kinds of equipment failures.   Finally as the cost of computation has decreased dramatically relative to the cost of communication it is now advantageous to trade off increased computation for reduced communication. Thus in database or sensor systems involving geographically separated data collection points it may be advantageous to process data locally at the point of collection and send condensed summaries to other points as needed rather than communicate the raw data to a single processing center.

For these reasons, we will be interested in schemes for distributed decision making and computation in which a set of processors (or decision makers) eventually compute a desired solution through a process of information exchange. It is possible to formulate mathematically a distributed decision problem whereby one tries to choose an "optimal" distributed scheme, subject to certain limitations. For example, we may impose constraints on the amount of information that may be transferred and look for a scheme which results in the best achievable decision, given these constraints. Such problems have been formulated and studied in the decentralized control context [21,22], as well as in the computer science literature [23,24]. However, in practice these turn out to be very difficult, usually intractable problems [25,26]. We therefore choose to focus on distributed algorithms with a prespecified structure (rather than try to find an optimal structure): we assume that each processor chooses an initial decision and iteratively improves this decision as more information is obtained from the environment or other processors. By this we mean that the ith processor updates from time to time his decision x^i using some formula

$$x^i \leftarrow f^i(x^i, I^i) \tag{1.1}$$

where I^i is the information available to the ith processor at the time of the update. In general there are serious limitations to this approach the most obvious of which is that the function f^i in (1.1) has to be chosen a priori on the basis of ad hoc considerations. However there are situations where the choice of reasonable functions f^i is not too difficult, and iterations such as (1.1) can provide a practical approach to an otherwise very difficult problem. After all, centralized counterparts of processes such as (1.1) are of basic importance in the study of stability of dynamic systems, and deterministic and stochastic optimization algorithms.

In most of the cases we consider the information I^i of processor i contains some past decisions of other processors. However, we allow the possibility that some processors perform computations (using (1.1)) more often than they exchange information, in which case the information I^i may be outdated. This allows us to model situations frequently encountered in large systems where it is difficult to maintain synchronization between various parts of the decision making and information gathering processes.

There are a number of characteristics and issues relating to the distributed iterative process (1.1) that either do not arise in connection with its centralized counterpart or else appear in milder form. First there is a graph structure characterizing the interprocessor flow of information. Second there is an expanded notion of the state of computation characterized by the current results of computation x^i and the latest information I^i available at the entire collection of processors i. Finally when (as we assume in this paper) there is no strict sequence according to which computation and communication takes place at the various processors the state of computation tends to evolve according to a point-to-set mapping and possibly in a probabilistic manner since each state of computation may give rise to many other states depending on which of the processors executes iteration (1.1) next and depending on possibly random exogenous information made available at the processors during execution of the algorithm.

From the point of view of applications, we can see several possible (broadly defined) areas. We discuss below some of them, although this is not meant to be an exhaustive list.

a) Parallel computing systems, possibly designed for a special purpose, e.g., for solving large scale mathematical programming problems with a particular structure. An important distinguishing feature of such systems is that the machine architecture is usually under the control of the designer. As mentioned above, we will assume a prespecified structure, thereby bypassing issues of architectural choice. However, the work surveyed in this paper can be useful for assessing the effects of communication delays and of the lack of synchronization in some parallel computing systems. Some of the early work on the subject [10], [11] is motivated by such systems. For a discussion of related issues see [7].

b) Data Communication Networks. Real time data network operation lends itself naturally to application of distributed algorithms. The structure needed for distributed computation (geographically distributed processors connected by communication links) is an inherent part of the system. Information such as link message flows, origin to destination data rates, and link and node failures is collected at geographically distributed points in the network. It is generally difficult to implement centralized algorithms whereby a single node would collect all information needed, make decisions, and transmit decisions back to the points of interest. The amount of data processing required of the central node may be too large. In addition the links over which information is transmitted to and from the central node are subject to failure thereby compounding the difficulties. For these reasons in many networks (e.g. the ARPANET) algorithms such as routing, flow control, and failure recovery are carried out in distributed fashion [1]-[5]. Since maintaining synchronization in a large data network generally poses implementation difficulties these algorithms are often operated asynchronously.

c) Distributed Sensor Networks and Signal Processing. Suppose that a set of sensors obtain noisy measurements (or a sequence of measurements) of a stochastic signal and then exchange messages with the purpose of computing a final estimate or identifying some unknown parameters. We are then interested in a scheme by which satisfactory estimates are produced without requiring that each sensor communicates his detailed information to a central processor. Some approaches that have been tried in this context may be found in [27,28,29,30].

d) Large Decentralized Systems and Organizations. There has been much interest, particularly in economics, in situations in which a set of rational decision makers make decisions and then update them on the basis of new information. Arrow and Hurwicz [31] have suggested a parallelism between the operation of an economic market and distributed computation. In this context the study of distributed algorithms may be viewed as an effort to model collective behavior. Similar models have bee proposed for biological systems [32]. Alternatively, finding good distributed algorithms and studying their communication requirements may yield insights on good ways of designing large organizations. It should be pointed out that there is an open debate concerning the degree of rationality that may be

assumed for human decision makers. Given the cognitive limitations of humans, it
is fair to say that only relatively simple algorithms can be meaningful in such
contexts. The algorithms considered in this paper tend to be simple particularly
when compared with other algorithms where decision makers attempt to process
optimally the available information.

There are several broad methodological issues associated with iterative
distributed algorithms such as correctness, computation or communication efficiency,
and robustness. In this paper we will focus on two issues that generally relate
to the question of validity of an algorithm.

a) Under what conditions is it possible to guarantee <u>asymptotic convergence</u>
of the iterates x^i for all processors i, and <u>asymptotic agreement</u> between different
processors i and j $[(x^i-x^j)\rightarrow 0]$?

b) How much <u>synchronization</u> between processor computations is needed in order
to guarantee asymptotic convergence or agreement?

Significant progress has been made recently towards understanding these issues
and the main purpose of this paper is to survey this work. On the other hand little
is known at present regarding issues such as speed of convergence, and assessment
of the value of communicated information in a distributed context. As a result we
will not touch upon these topics in the present paper. Moreover, there are certain
settings (e.g., decentralized control of dynamical systems, dynamic routing in data
networks) in which issues of asymptotic convergence and agreement do not arise.
Consequently, the work surveyed here is not of direct relevance to such situations.

In the next two sections we formulate a model of distributed asynchronous
iterative computation, and illustrate its relevance by means of a variety of examples
from optimization, parameter estimation, and communication networks. The model
bears similarity to models of chaotic relaxation and distributed asynchronous fixed
point computation [10]-[13] but is more general in two respects. First we allow
two or more processors to update separately estimates of the same coordinate of
the decision vector and combine their individual estimates by taking convex
combinations, or otherwise. Second we allow processors to receive possibly stochastic measurements from the environment which may depend in nonlinear fashion on
estimates of other processors. These generalizations broaden a great deal the
range of applicability of the model over earlier formulations.

In Sections 4 and 5 we discuss two distinct approaches for analyzing algorithmic convergence. The first approach is essentially a generalization of the
Lyapounov function method for proving convergence of centralized iterative processes.
The second approach is based on the idea that if the processors communicate fast
relative to the speed of convergence of computation then their solution estimates
will be close to the path of a certain centralized process. By analyzing the convergence of this latter process one can draw inferences about the convergence of
the distributed process. In Section 5 we present results related primarily to

deterministic and stochastic descent optimization algorithms. An analysis that
parallels Ljung's ODE approach [37], [38] to recursive stochastic algorithms may be
found in [35] and in a forthcoming publication. In Section 6 we discuss convergence
and agreement results for a special class of distributed processes in which the
update of each processor, at any given time, is the optimal estimate of a solution
given his information, in the sense that it minimizes the conditional expectation
of a common cost function.

2. A Distributed Iterative Computation Model

In our model we are given a set of feasible decisions X and we are interested in finding an element of a special subset X* called the solution set. We do not specify X* further for the time being. An element of X* will be referred to as a solution. Without loss of generality we index all events of interest (message transmissions and receptions, obtaining measurements, performing computations) by an integer time variable t. There is a finite collection of processors $i=1,\ldots,n$ each of which maintains an estimate $x^i(t) \in X$ of a solution and updates it once in a while according to a scheme to be described shortly. The ith processor receives also from time to time m_i different types of measurements and maintains the latest values $z_1^i, z_2^i, \ldots, z_{m_i}^i$ of these measurements. (That is, if no measurement of type j is received at time t, then $z_j^i(t+1) = z_j^i(t)$). The measurement z_j^i is an element of a set Z_j^i. Each time a measurement z_j^i of type j is received by processor i the old value z_j^i is replaced by the new value and the estimate x^i is updated according to

$$x^i(t+1) = M_{ij}(x^i(t), z_1^i(t), \ldots, z_{m_j}^i(t)) , \qquad (2.1)$$

where M_{ij} is a given function. Each node i also updates from time to time the estimate x^i according to

$$x^i(t+1) = C_i(x^i(t), z_1^i(t), \ldots, z_{m_i}^i(t)) \qquad (2.2)$$

where C_i is a given function. Thus at each time t each processor i either receives a new measurement of type j and updates x^i according to (2.1), or updates x^i according to (2.2), or remains idle in which case $x^i(t+1) = x^i(t)$ and $z_j^i(t+1) = z_j^i(t)$ for all j. The sequence according to which a processor executes (2.1) or (2.2) or remains idle is left unspecified and indeed much of the analysis in this paper is oriented towards the case where there is considerable a priori uncertainty regarding this sequence. One of the advantages of this approach is that difficult analytical problems arising due to consideration of non-classical information patterns [21] do not appear in our framework. Note that neither mapping M_{ij} or C_i involves a dependence on the time argument t. This is appropriate since it would be too restrictive to assume that all processors have access to a global clock that records the current time index t. On the other hand the mappings M_{ij} and C_i may include dependences on local clocks (or counters) that record the number of times iterations (2.1) or (2.2) are executed at processor i. The value of the local counter of processor i may be artificially lumped as an additional component into the estimate x^i and incremented each time (2.1) or (2.2) are executed.

Note that there is redundancy in introducing the update formula (2.2) in addition to (2.1). We could view (2.2) as a special case of (2.1) corresponding to an update in response to a "self-generated" measurement at node i. Indeed such a formulation may be appropriate in some problems. On the other hand there is often some conceptual value in separating the types of updates at a processor in updates that incorporate new exogenous information (cf. (2.1)), and updates that utilize the existing information to improve the processor's estimate (cf. (2.2)).

The measurement $z_j^i(t)$, received by processor i at time t, is related to the processor estimates $x^1, x^2, \ldots, x^n$ according to an equation of the form

$$z_j^i(t) = \phi_{ij}(x^1(\tau_j^{i1}(t)), x^2(\tau_j^{i2}(t)), \ldots x^n(\tau_j^{in}(t)), \omega),$$ (2.3)

where ω belongs to the sample space Ω corresponding to a probability space (Ω, F, P).

We allow the presence of delays in equation (2.3) in the sense that the estimates $x^1, \ldots, x^n$ may be the ones generated via (2.1) or (2.2) at the corresponding processors at some times $\tau_j^{ik}(t) \leq t$, prior to the time t that $z_j^i(t)$ was received at processor i. Furthermore the delays may be different for different processors. We place the following restriction on these delays which essentially says that successive measurements of the same type depend on successive processor estimates.

Assumption 2.1: If $t \geq t'$, then

$$\tau_j^{ik}(t) \geq \tau_j^{ik}(t'), \quad \forall i, j, k .$$

For the time being, the only other assumption regarding the timing, and sequencing of measurement reception and estimate generation is the following:

Assumption 2.2 (Continuing Update Assumption): For any i and j and any time t there exists a time t'>t at which a measurement z_j^i of the form (2.3) will be received at i and the estimate x^i will be updated according to (2.1). Also for any i and time t there exists a time t">t at which the estimate x^i will be updated according to (2.2).

The assumption essentially states that each processor will continue to receive measurements in the future and update his estimate according to (2.1) and (2.2). Given that we are interested in asymptotic results there isn't much we can hope to prove without an assumption of this type. In order to formulate substantive convergence results we will also need further assumptions on the nature of the mappings M_{ij}, C_{ij}, and ϕ_{ij} and possibly on the relative timing of measurement receptions, estimate updates and delays in (2.3) and these will be introduced later. In the next section we illustrate the model and its potential uses by means of examples.

It should be pointed out here that the above model is very broad and may capture a large variety of different situations, provided that the measurements z_j^i are given appropriate interpretations. For example, the choice $z_j^i(t) = x^j(\tau_j^{ij}(t))$

corresponds to a situation where processor i receives a message with the estimate computed by processor j at time $\tau_j^{ij}(t)$, and $t-\tau_j^{ij}(t)$ may be viewed as a communication delay. In this case processors act also as sensors generating measurements for other processors. In other situations however specialized sensors may generate (possibly noisy and delayed) feedback to the processors regarding estimates of other processors of (cf. (2.3)). Examples of both of these situations will be given in the next section.

3. Examples

An important special case of the model of the previous section is when the feasible set X is the Cartesian product of n sets

$$X = X_1 \times X_2 \times \ldots \times X_n,$$

each processor i is assigned the responsibility of updating the ith component of the decision vector $x = (x_1, x_2, \ldots, x_n)$ via (2.1) or (2.2) while receiving from each processor j ($j \neq i$) the value of the jth component x_j. We refer to such distributed processes as being __specialized__. The first five examples are of this type.

Example 1: (Shortest Path Computation)

Let (N, A) be a directed graph with set of nodes $N = \{1, 2, \ldots, n\}$ and set of links A. Let $N(j)$ denote the set of downstream neighbors of node i, i.e. the nodes j such that (i,j) is a link. Assume that each link (i,j) is assigned a positive scalar a_{ij} referred to as its length. Assume also that there is a directed path to node 1 from every other node. Let x_i^i be the estimate of the shortest distance from node i to node 1 available at node i. Consider a distributed algorithm whereby each node $i=1, \ldots, n$ executes the iteration

$$x_i^i \leftarrow \min_{j \in N(i)} \{a_{ij} + x_j^j\} \qquad (3.1)$$

after receiving one or more estimates x_j^j from its neighbors, while node 1 sets

$$x_1^1 = 0.$$

This algorithm—a distributed asynchronous implementation of Bellman's shortest path algorithm—was implemented on the ARPANET in 1969 [14]. The estimate x_i^i can be shown to converge to the unique shortest distance from node i to node 1 provided the starting values x_i^i are nonnegative [12]. The algorithm clearly is a special case of the model of the previous section. Here the measurement equation [cf. (2.3)] is

$$z_j^i = x_j^j, \qquad \forall\ j \in N(i) \qquad (3.2)$$

the measurement update equation [cf. (2.1)] replaces x_j^i by z_j^i and leaves all other coordinates x_m^i, $m \neq j$ unchanged, while the corresponding update formula of (2.2) can be easily constructed using (3.1).

Example 2: (Fixed point calculations)

The preceding example is a special case of a distributed dynamic programming algorithm (see [12]) which is itself a special case of a distributed fixed point algorithm. Suppose we are interested in computing a fixed point of a mapping

F: $X \to X$. We construct a distributed fixed point algorithm that is a special case of the model of the previous section as follows:

Let X be a Cartesian product of the form $X = X_1 \times X_2 \times \ldots \times X_n$ and let us write accordingly $x = (x_1, x_2, \ldots, x_n)$ and $F(x) = (F_1(x), F_2(x), \ldots, F_n(x))$ where $F_i : X \to X_i$. Let $x^i = (x_1^i, \ldots, x_n^i)$ be the estimate of x generated at the ith processor. Processor i executes the iteration

$$x_j^i \leftarrow \begin{cases} x_j^i & \text{if } i \neq j \\ F_i(x^i) & \text{if } i = j, \end{cases} \qquad (3.3)$$

(this corresponds to the mapping C_i of (2.2)), and transmits from time to time x_i^i to the other processors. Thus the measurements z_j^i are given by [cf. (2.3)]

$$z_j^i = x_j^j, \quad i \neq j \qquad (3.4)$$

and the (i,j)th measurement update equation [cf. (2.1)] is given by

$$x_m^i \leftarrow \begin{cases} x_m^i & \text{if } m \neq j \\ z_j^i & \text{if } m = j. \end{cases} \qquad (3.5)$$

Conditions under which the estimate x^i converges to a fixed point of F are given in [13] (see also Section 4).

Example 3: (Distributed deterministic gradient algorithm)

This example is a special case of the preceding one whereby $X = R^n$, $X_i = R$, and F is of the form

$$F(x) = x - \alpha \nabla f(x) \qquad (3.6)$$

where ∇f is the gradient of a function $f: R^n \to R$, and α is a positive scalar stepsize. Iteration (3.3) can then be written as

$$x_j^i \leftarrow \begin{cases} x_j^i & \text{if } i \neq j \\ x_i^i - \alpha \dfrac{\partial f(x^i)}{\partial x_i} & \text{if } i = j \end{cases} \qquad (3.7)$$

A variation of this example is obtained if we assume that, instead of each processor i transmitting directly his current value of the coordinate x_i to the other processors, there is a measurement device that transmits the current value of the partial derivative $\dfrac{\partial f(x)}{\partial x_i}$ to the ith processor. In this case there is only one type of measurement for each processor i [cf. (2.3)] and it is given by

$$z_1^i = \frac{\partial f(x_1^1, \ldots, x_n^n)}{\partial x_i} .$$

While the equation above assumes no noise in the measurement of each partial derivative one could also consider the situation where this measurement is corrupted by
additive or multiplicative noise thereby obtaining a model of a distributed stochastic gradient method. Many other descent algorithms admit a similar distributed
version.

Example 4: (An Organizational Model)

This example is a variation of the previous one, but may be also viewed as a
model of collective decision making in a large organization. Let $X = X_1 \times X_2 \times \ldots \times X_n$
be the feasible set, where X_i is a Euclidean space and let $f: X \to [0,\infty)$ be a cost
function of the form $f(x) = \sum_{i=1}^{n} f^i(x)$. We interpret f^i as the cost facing the
i-th division of an organization. This division is under the authority of decision
maker (processor) i, who updates the i-th component $x_i \in X_i$ of the decision vector
x. We allow the cost f^i to depend on the decisions x_j of the remaining decision
makers, but we assume that this dependence is weak. That is, let

$$K_{jm}^i = \sup_{x \in X} \left| \frac{\partial^2 f^i(x)}{\partial x_j \partial x_m} \right|$$

and we are interested in the case $K_{jm}^i \ll K_{ii}^i$ (unless j=m=i). Decision maker i
receives measurements z_j^i, j=1,...,n of the form

$$z_j^i(t) = \frac{\partial f^j}{\partial x_i} (x_1^1(\tau_j^{i1}(t)), x_2^2(\tau_j^{i2}(t)), \ldots, x_n^n(\tau_j^{in}(t))), \qquad (3.8)$$

where $\tau_j^{ik}(t) \le t$ [cf. (2.3)]. Once in a while, he also updates his decision according to

$$x_i^i(t+1) = x_i^i(t) - \alpha_i \sum_{j=1}^{n} z_j^i(t) . \qquad (3.9)$$

If we assume that

$$\tau_j^{im}(t) \le \tau_j^{ij}(t), \qquad \forall i,j,m,t,$$

the above algorithm admits the following interpretation: each decision maker m, at
time $\tau_j^{im}(t)$ sends a message $x_m^m(\tau_j^{im}(t))$, to inform decision maker j of his decision.
These messages are the last such messages received by decision maker j no later
than $\tau_j^{ij}(t)$. Then, decision maker j (who is assumed to be knowledgeable about f^j)

computes z_j^i according to (3.8) and sends it to decision maker i; the latter message is the last such message received by decision maker i no later than t, and is being used, at time t, by decision maker i, to update his decision according to (3.9). On an abstract level, each decision maker j is being informed about the decision of the others and replies by saying how he is affected by their decisions; however, this may be done in an asynchronous and very irregular manner.

Example 5:   (Distributed optimal routing in data networks)

A standard model of optimal routing in data networks (see e.g. the survey [6]) involves the multicommodity flow problem

$$\text{minimize } \sum_{a \in A} D_a(F_a)$$

$$\text{subject to } F_a = \sum_{w \in W} \sum_{\substack{p \in P_w \\ a \in p}} x_p, \quad \forall \ a \in A$$

$$\sum_{p \in P_w} x_p = r_w, \quad \forall \ w \in W$$

$$x_p \geq 0, \quad \forall \ w \in W, \ p \in P_w.$$

Here A is the set of directed links in a data network, F_a is the communication rate (say in bits/sec) on link $a \in A$, W is a set of origin-destination (OD) pairs, P_w is a given set of directed paths joining the origin and the destination of OD pair w, x_p, $p \in P_w$ is the communication rate on path p of OD pair w, r_w is a given required communication rate of OD pair w, and D_a is a monotonically increasing differentiable convex function for each $a \in S$.   The objective here is to distribute the required rates r_w among the available paths in P_w so as to minimize a measure of average delay per message as expressed by $\sum_{a \in A} D_a(F_a)$.

Since the origin node of each OD pair w has control over the rates x_p, $p \in P_w$ it is convenient to use a distributed algorithm of the gradient projection type (see [6], [8]) whereby each origin iterates on its own path rates asynchronously and independently of other origins.   This type of iteration requires knowledge of the first partial derivatives $D_a'(F_a)$ for each link evaluated at the current link rates F_a.   A practical scheme similar to the one currently adopted on the ARPANET [9] is for each link $a \in A$ to broadcast to all the nodes the current value of either F_a or $D_a'(F_a)$.   This information is then incorporated in the gradient projection iteration of the origin nodes.   In this scheme each origin node can be viewed as a processor and F_a or $D_a'(F_a)$ plays the role of a measurement which depends on the solution estimates of all processors [cf. (2.3)].

The direct opposite of a specialized process, in terms of division of labor between processors is a totally overlapping process.

Example 6:   (Total Overlap)

Let the feasible set X be a Euclidean space. Each processor i receives measurements z_j^i ($j\neq i$) which are the values of the estimates x^j of other processors; that is,

$$z_j^i = x^j, \quad i \neq j.$$

Whenever such a measurement is received, processor i updates his estimate by taking a convex combination:

$$x^i \leftarrow M_{ij}(x^i, z_j^i) = \beta^{ij}x^i + (1-\beta^{ij}) z_j^i \tag{3.10}$$

where $0 < \beta^{ij} < 1$. Also processor i receives his own information z_i^i, generated according to

$$z_i^i = \phi_i(x^i, \omega)$$

and updates x^i according to

$$x^i(t+1) = M_{ii}(x^i(t), z_i^i(t)) = x^i(t) - \alpha^i z_i^i(t) \tag{3.11}$$

where α^i is a positive scalar stepsize.[†] Such an algorithm is of interest if the objective is to minimize a cost function $f: X \to \mathbb{R}$, and $z_i^i(t)$ is in some sense a descent direction with respect to f. In a deterministic setting, such a scheme could be redundant, as some processors would be close to replicating the computation of others. In a stochastic setting, however (e.g. if

$$z_i^i(t) = \frac{\partial f}{\partial x}(x^i(t)) + w^i(t),$$

where $w^i(t)$ is zero-mean white noise) the combining process is effectively averaging out the effects of the noise and may improve convergence.

Example 7:   (System Identification)

Consider two moving average processes $y^1(t), y^2(t)$ generated according to

---

[†] The stepsice α^i could be constant as in deterministic gradient methods. However, in other cases (such as stochastic gradient methods with additive noise) it is essential that α^i is time varying and tends to zero. This, strictly speaking, violates the assumption that the mapping M_{ij} does not depend on the time t. However it is possible to circumvent this by introducing (as an additional component of x^i) a local counter as each processor i that keeps track of the number of times iteration (3.10) or (3.11) is executed at processor i. The stepsizé α^i could be made dependent on the value of this local counter (see the discussion following (2.1) and (2.2) in Section 2).

$$y^i(t) = A(q)u(t) + w^i(t),$$

where $A(\cdot)$ is a polynomial, to be identified, q is the unit delay operator and $w^i(t)$, $i=1,2$, are white, zero-mean processes, possibly correlated with each other. Let there be two processors ($n=2$); processor i measures $y^i(t)$ and both measure $u(t)$ at each time t. Each processor i updates his estimate x^i of the coefficients of A according to any of the standard system identification algorithms (e.g. the LMS or RLS algorithm). Under the usual identifiability conditions [33] each processor would be able to identify $A(\cdot)$ by itself. However, convergence should be faster if once in a while one processor gets (possibly delayed) measurements of the estimates of the other processor and combines them by taking a convex combination. Clearly, this is a special case of Example 6.

A more complex situation arises if we have two ARMAX processes y^1, y^2, driven by a common colored noise:

$$A^i(q)y^i(t) = B^i(q)u^i(t) + w(t), \quad i=1,2$$

$$w(t) = C(q)v(t),$$

where $v(t)$ is white and A^i, B^i, C are polynomials in the delay operator q. Assuming that each processor i observes y^i and u^i, he may under certain conditions [34] identify A^i, B^i. In doing this he must, however, identify the common noise source C as well. So we may envisage a scheme whereby processors use a standard algorithm to identify A^i, B^i, C and once in a while exchange messages with their estimates of the coefficients of C; these estimates are then combined by taking a convex combination.

This latter example falls in between the extreme cases of specialization and total overlap: there is specialization concerning the coefficients of A^i, B^i and overlap concerning the coefficients of C.

4. Convergence of Contracting Processes

In our effort to develop a general convergence result for the distributed algorithmic model of Section 2 we draw motivation from existing convergence theories for (centralized) iterative algorithms. There are several theories of this type (Zangwill [15], Luenberger [16], Ortega and Rheinboldt [17]--the most general are due to Poljak [18] and Polak [19]). Most of these theories have their origin in Lyapunov's stability theory for differential and difference equations. The main idea is to consider a generalized distance function (or Lyapunov function) of the typical iterate to the solution set. In optimization methods the objective function is often suitable for this purpose while in equation solving methods a norm of the difference between the current iterate and the solution is usually employed. The idea is typically to show that at each iteration the value of the distance function is reduced and reaches its minimum value in the limit.

The result of this section is based on a similar idea. However instead of working with a generalized distance function we prefer to work (essentially) with the level sets of such a function; and instead of working with a single processor iterate (as in centralized processes) we work with what may be viewed as a state of computation of the distributed processes which includes all current processor iterates and all latest information available at the processors.

The subsequent result is reminiscent of convergence results for successive approximation methods associated with contraction mappings. For this reason we refer to processes satisfying the following assumption as <u>contracting processes</u>. In what follows in this section we assume that the feasible set X in the model of Section 2 is a topological space so we cant talk about convergence of sequences in X.

<u>Assumption 3.1</u>: There exists a sequence of sets $X(k)$ with the following properties:

a) $X^* \subset X(k+1) \subset X(k) \subset \ldots \subset X$.

b) If $\{x_k\}$ is a sequence in X such that $x_k \in X(k)$ for all k, then $\{x_k\}$ converges to a solution.

(<u>Note</u>: If the notion of sequence convergence to a subset is defined on X, one may replace convergence of $\{x_k\}$ to a solution with convergence to the solution set X^*).

c) For all i, j and k denote:

$$z_j^i(k) = \{\phi_{ij}(x^1,\ldots,x^n,\omega) \mid x^1 \in X(k) \ldots,x^n \in X(k), \omega \in \Omega\} \tag{4.1}$$

$$\overline{x}^i(k) = \{C_i(x^i,z_1^i,\ldots,z_{m_i}^i) \mid x^i \in X(k), z_1^i \in z_1^i(k),\ldots,z_{m_i}^i \in z_{m_i}^i(k)\} \tag{4.2}$$

$$\overline{z}_j^i(k) = \{\phi_{ij}(x^1,\ldots,x^n,\omega) \mid x^1 \in \overline{x}^1(k),\ldots,x^n \in \overline{x}^n(k), \omega \in \Omega\} \tag{4.3}$$

The sets $X(k)$ and the mappings ϕ_{ij}, M_{ij}, and C_i are such that for all i,j and k

$$\bar{x}^i(k) \subset X(k) \tag{4.4}$$

$$M_{ij}(x^i, z^i_1, \ldots, z^i_{m_i}) \in X(k), \quad \forall x^i \in X(k), \quad z^i_1 \in Z^i_1(k), \ldots, z^i_{m_i} \in Z^i_{m_i}(k) \tag{4.5}$$

$$M_{ij}(x^i, z^i_1, \ldots, z^i_{m_i}) \in \bar{x}^i(k), \quad \forall x^i \in \bar{x}^i(k), \quad z^i_1 \in Z^i_1(k), \ldots, z^i_{m_i} \in Z^i_{m_i}(k) \tag{4.6}$$

$$M_{ij}(x^i, z^i_1, \ldots, z^i_{m_i}) \in X(k+1), \quad \forall x^i \in \bar{x}^i(k), \quad z^i_1 \in \bar{Z}^i_1(k), \ldots, z^i_{m_i} \in \bar{Z}^i_{m_i}(k). \tag{4.7}$$

Assumption 3.1 is a generalized version of a similar assumption in reference [13]. Broad classes of deterministic specialized processes satisfying the assumption are given in that reference. The main idea is that membership in the set $X(k)$ is representative in some sense of the proximity of a processor estimate to a solution. By part b), if we can show that a processor estimate successively moves from $X(0)$ to $X(1)$, then to $X(2)$ and so on, then convergence to a solution is guaranteed. Part c) assures us that once all processor estimates enter the set $X(k)$ then they remain in the set $X(k)$ [cf. (4.4), (4.5)] and (assuming all processors keep on computing and receiving measurements) eventually enter the set $X(k+1)$ [cf. (4.6), (4.7)]. In view of these remarks the proof of the following result is rather easy. Note that the assumption does not differentiate the effects of two different members of the probability space [cf. part c)] so it applies to situations where the process is either deterministic (Ω consists of a single element), or else stochastic variations are not sufficiently pronounced to affect the membership relations in part c).

Proposition 3.1: Let Assumptions 2.1, 2.2., 3.1, hold and assume that all initial processor estimates x^i, $i=1,\ldots,n$ belong to $X(0)$, while all initial measurements z^i_j available at the processors belong to the corresponding sets $Z^i_j(0)$. Then each of the sequences $\{x^i(t)\}$ converges almost surely to a solution as $t \to \infty$.

The proof will not be given since it is very similar to the one given in [13]. Note that the proposition does not guarantee asymptotic agreement of the processor estimates but in situations where Assumption 3.1 is satisfied one can typically also show agreement.

Example 2 (continued): As an illustration consider the specialized process for computing a fixed point of a mapping F in example 2. There X is a Cartesian product $X_1 \times X_2 \times \ldots \times X_n$, and each processor i is responsible for updating the ith "coordinate" x_i of $x = (x_1, x_2, \ldots, x_n)$ while relying on essentially direct communications from other processors to obtain estimates of the other coordinates. Suppose that each set X_i is a Banach space with norm $||\cdot||_i$ and X is endowed with the sup norm

$$||x|| = \max\{||x_1||_1, \ldots, ||x_n||_n\}, \qquad \forall x \in X \tag{4.8}$$

Assume further that F is a contraction mapping with respect to this norm, i.e. for some $\alpha \in (0,1)$

$$||F(x)-F(y)|| \leq \alpha ||x-y||, \qquad \forall x,y \in X. \tag{4.9}$$

Then the solution set consists of the unique fixed point x* of F. For some positive constant B let us consider the sequence of sets

$$X(k) = \{x \in X | \ ||x-x^*|| \leq B\alpha^k\}, \qquad k = 0,1,\ldots,$$

The sets defined by (4.1)-(4.3) are then given by

$$z_j^i(k) = \{x_j \in X_j | \ ||x_j-x_j^*||_j \leq B\alpha^k\}$$

$$\bar{x}^i(k) = \{x \in X(k)| \ ||x_i-x_i^*||_i \leq B\alpha^{k+1}\}$$

$$\bar{z}_j^i(k) = \{x_j \in X_j | \ ||x_j-x_j^*||_j \leq B\alpha^{k+1}\}.$$

It is straightforward to show that the sequence $\{X(k)\}$ satisfies Assumption 3.1. Further illustrations related to this example are given in [13]. Note however that the use of the sup norm (4.8) is essential for the verification of Assumption 3.

Similarly Assumption 3 can be verified in the preceding example if the contraction assumption (4.9) is substituted by a monotonicity assumption (see [13]). This monotonicity assumption is satisfied by most of the dynamic programming problems of interest including the shortest path problem of example 1 (see also [12]). An important exception is the infinite horizon average cost Markovian decision problem (see [12] , p. 616).

An important special case for which the contraction mapping assumption (4.9) is satisfied arises when $X=R^n$ and $x_1,x_2,\ldots,x_n$ are the coordinates of x. Suppose that F satisfies

$$|F(x)-F(y)| \leq P|x-y|, \qquad \forall x,y \in R^n \tag{4.10}$$

where P is an nxn matrix with nonnegative elements and spectral radius strictly less than unity, and for any $z=(z_1,z_2,\ldots,z_n)$ we denote by $|z|$ the column vector with coordinates $|z_1|,|z_2|,\ldots,|z_n|$. Then F is called a P-contraction mapping. Fixed point problems involving such mappings arise in dynamic programming ([20] , p.374), and solution of systems of nonlinear equations ([17], Section 13.1). It can be shown ([11], p.231) that if F is a P-contraction then it is a contraction mapping with respect to some norm of the form (4.8). Therefore Proposition 3.1 applies.

We finally note that it is possible to use Proposition 3.1 to show convergence of similar fixed point distributed processes involving partial or total overlap between the processors (compare with example 6).

Example 3 (continued): Consider the special case of the deterministic gradient algorithm of example 3 corresponding to the mapping

$$F(x) \; = \; x - \alpha \nabla f(x). \tag{4.11}$$

Assume that $f:R^n \rightarrow R$ is a twice continuously differentiable convex function with Hessian matrix $\nabla^2 f(x)$ which is positive definite for all x. Assume also that there exists a unique minimizing point x* of f over R^n. Consider the matrix

$$H^* \; = \; \begin{bmatrix} \left| \dfrac{\partial^2 f(x^*)}{(\partial x_1)^2} \right| , & - \left| \dfrac{\partial^2 f(x^*)}{\partial x_1\,\partial x_2} \right| & ,\ldots, & - \left| \dfrac{\partial^2 f(x^*)}{\partial x_1\,\partial x_n} \right| \\ \vdots & \vdots & & \vdots \\ \left| \dfrac{\partial^2 f(x^*)}{\partial x_n\,\partial x_1} \right| , & - \left| \dfrac{\partial^2 f(x^*)}{\partial x_n\,\partial x_2} \right| & ,\ldots, & \left| \dfrac{\partial^2 f(x^*)}{(\partial x_n)^2} \right| \end{bmatrix} \tag{4.12}$$

obtained from the Hessian matrix $\nabla^2 f(x^*)$ by replacing the off-diagonal terms by their negative absolute values. It is shown in [13] that if the matrix H* is positive definite then the mapping F of (4.11) is a P-contraction within some open sphere centered at x* provided the stepsize α in (4.11) is sufficiently small. Under these circumstances the distributed asynchronous gradient method of this example is convergent to x* provided all initial processor estimates are sufficient-ly close to x* and the stepsize α is sufficiently small. The neighborhood of local convergence will be larger if the matrix (4.12) is positive definite within an ac-cordingly larger neighborhood of x*. For example if f is positive definite quadratic with the corresponding matrix (4.12) positive definite a global convergence result can be shown.

One condition that guarantees that H* is positive definite is strict diagonal dominance ([17], p.48-51).

$$\frac{\partial^2 f}{(\partial x_i)^2} \; > \; \sum_{\substack{j=1 \\ j \neq i}}^{n} \left| \frac{\partial^2 f}{\partial x_i\,\partial x_j} \right| , \quad \forall i=1,\ldots,n,$$

where the derivatives above are evaluated at x*. This type of condition is typical-ly associated with situations where the coordinates of x are weakly coupled in the sense that changes in one coordinate have small effects on the first partial deriv-atives of f with respect to the other coordinates. This result can be generalized to the case of weakly coupled systems (as opposed to weakly coupled coordinates).

Assume that x is partitioned as $x = (x_1, x_2, \ldots, x_n)$ where now $x_i \in R^{m_i}$ (m_i may be greater than one but all other assumptions made earlier regarding f are in effect). Assume that there are n processors and the ith processor asynchronously updates the subvector x_i according to an approximate form of Newton's method where the second order submatrices of the Hessian $\nabla^2_{x_i x_j} f$, $i \neq j$ are neglected, i.e.

$$x_i \leftarrow x_i - (\nabla^2_{x_i x_i} f)^{-1} \nabla_{x_i} f. \tag{4.13}$$

In calculating the partial derivatives above processor i uses the values x_j latest communicated from the other processors $j \neq i$ similarly as in the distributed gradient method. It can be shown that if the cross-Hessians $\nabla^2_{x_i x_j} f$, $i \neq j$ have sufficiently small norm relative to $\nabla^2_{x_i x_i} f$, then the totally asynchronous version of the approximate Newton method (4.13) converges to x* if all initial processor estimates are sufficiently close to x*. The same type of result may also be shown if (4.13) is replaced by

$$x_i \leftarrow \arg \min_{x_i \in R^{m_i}} f(x_1, x_2, \ldots, x_n). \tag{4.14}$$

Unfortunately it is not true always that the matrix (4.12) is positive definite, and there are problems where the totally asynchronous version of the distributed gradient method is not guaranteed to converge regardless of how small the stepsize α is chosen. As an example consider the function $f: R^3 \to R$

$$f(x_1, x_2, x_3) = (x_1 + x_2 + x_3)^2 + (x_1 + x_2 + x_3 - 3)^2 + \varepsilon (x_1^2 + x_2^2 + x_3^2)$$

where $0 < \varepsilon << 1$. The optimal solution is close to $(\frac{1}{2}, \frac{1}{2}, \frac{1}{2})$ for ε: small. The scalar ε plays no essential role in this example. It is introduced merely for the purpose of making the Hessian of f positive definite. Assume that all initial processor estimates are equal to some common value $\bar{x}$, and that processors execute many gradient iterations with a small stepsize before communicating the current values of their respective coordinates to other processors. Then (neglecting the terms that depend on ε) the ith processor tries in effect to solve the problem

$$\min_{x_i} \{ (x_i + 2\bar{x})^2 + (x_i + 2\bar{x} - 3)^2 \}$$

thereby obtaining a value close to $\frac{3}{2} - 2\bar{x}$. After the processor estimates of the respective coordinates are exchanged each processor coordinate will have been updated approximately according to

$$\bar{x} \leftarrow \frac{3}{2} - 2\bar{x} \tag{4.15}$$

and the process will be repeated. Since (4.15) is a divergent iterative process we see that, regardless of the stepsize chosen and the proximity of the initial processor estimates to the optimal solution, by choosing the communication delays sufficiently large the distributed gradient method can be made to diverge when the matrix H* of (4.12) is not positive definite.

5.   Convergence of Descent Processes

We saw in the last section that the distributed gradient algorithm converges
appropriately when the matrix (4.12) is positive definite. This assumption is not
always satisfied, but convergence can be still shown (for a far wider class of
algorithms) if a few additional conditions are imposed on the frequency of obtaining
measurements and on the magnitude of the delays in equation (2.3). The main idea
behind the results described in this section is that if delays are not too large,
if certain processors do not obtain measurements and do not update much more
frequently than others, then the effects of asynchronism are relatively small and
the algorithm behaves approximately as a centralized algorithm, similar to the class
of centralized pseudogradient algorithms considered in [40].

Let $X = X_1 \times X_2 \times \ldots \times X_L$ be the feasible set, where X_ℓ, $\ell=1,\ldots,L$, is a reflexive
Banach space. If $x = (x_1,\ldots,x_L)$, $x_\ell \in X_\ell$, we refer to x_ℓ as the ℓ-th component of x.
We endow X with the sup norm, as in (4.8). Let $f: X \rightarrow [0,\infty)$ be a cost function to be
minimized. We assume that f is Frechet differentiable and its derivative is Lipschitz
continuous.

Each processor i keeps in its memory an estimate $x^i(t) = (x_1^i(t),\ldots,x_L^i(t)) \in X$
and receives measurements $z_{j,\ell}^i \in X_\ell$, $i \neq j$, with the value of the ℓ-th component of
x^j,   evaluated by processor j at some earlier time $\tau_{j,\ell}^i(t) \leq t$; that is,
$z_{j,\ell}^i(t) = x_\ell^j(\tau_{j,\ell}^i(t))$. He also receives from the environment exogenous, possibly
stochastic measurements $z_i^i \in X$, which are in a direction of descent with respect
to the cost function f, in a sense to be made precise later. We denote by $z_{i,\ell}^i$ the
ℓ-th component of z_i^i.

Whenever processor i receives measurements $z_{j,\ell}^i$, he updates his estimate vector
x^i componentwise, according to:

$$x_\ell^i(t+1) = \beta_{i,\ell}^i(t) x_\ell^i(t) + \sum_{j \neq i} \beta_{j,\ell}^i(t) z_{j,\ell}^i(t) + \alpha^i(t) z_{i,\ell}^i(t). \qquad (5.1)$$

The coefficients $\beta_{j,\ell}^i(t)$ are nonegative scalars satisfying

$$\sum_{j=1}^{n} \beta_{j,\ell}^i(t) = 1, \qquad \forall i,\ell,t,$$

and such that: if no measurement $z_{j,\ell}^i$ was received by processor i ($i \neq j$) at time t,
then $\beta_{j,\ell}^i(t) = 0$. That is, processor i combines his estimate of the ℓ-th component
of the solution with the estimates (possibly outdated) of other processors that he
has just received, by forming a convex combination. Also, if no new measurement z_i^i
was obtained at time t, we should set $z_i^i(t) = 0$ in equation (5.1). The coefficient
$\alpha^i(t)$ is a nonegative stepsize. It can be either independent of t or it may depend
on the number of times up to t that a new measurement (of any type) was received
at processor i.

Equation (5.1) which essentially defines the algorithm, is a linear system driven by the exogenous measurements $z_i^i(t)$. Therefore, there exist linear operators $\phi^{ij}(t|s)$, $(t \geq s)$, such that

$$x^i(t) = \sum_{j=1}^{n} \phi^{ij}(t|0)x^j(1) + \sum_{s=1}^{t-1} \sum_{j=1}^{n} \alpha^j(s)\phi^{ij}(t|s)z_j^j(s).$$

We now impose an assumption which states that if the processors cease obtaining exogenous measurements from some time on (that is, if they set $z_i^i=0$), they will asymptotically agree on a common limit:

Assumption 5.1: For any i,j,s, $\lim_{t\to\infty} \phi^{ij}(t|s)$ exists (with respect to the induced operator norm) and is the same for all i. The common limit is denoted by $\phi^j(s)$.

Assumption 5.1 is very weak. Roughly speaking it requires that for every component $\ell \in \{1,\ldots,L\}$ there exists a directed graph $G=(N,A)$, where the set N of nodes is the set $\{1,\ldots,n\}$ of processors such that there exists a path from every processor to every other processor, and such that if (i,j) belongs to A, then processor j receives an infinite number of measurements of type $z_{i,\ell}^j$. Also the coefficients $\beta_{i,\ell}^j(t)$ must be such that "sufficient combining" takes place and the processors tend to agree.

We can now define a vector $y(t) \in X$ by

$$y(t) = \sum_{j=1}^{n} \phi^j(0)x^j(1) + \sum_{s=1}^{t-1} \sum_{j=1}^{n} \alpha^j(s)\phi^j(s)z_j^j(s)$$

and observe that $y(t)$ is recursively generated by

$$y(t+1) = y(t) + \sum_{i=1}^{n} \alpha^i(t)\phi^i(t)z_i^i(t). \tag{5.2}$$

We can now explain the main idea behind the results to be described: if $\phi^{ij}(t|s)$ converges to $\phi^j(s)$ fast enough, if $\alpha^i(t)$ is small enough, and if $z_i^i(t)$ is not too large, then $x^i(t)$, for each i, will evolve approximately as $y(t)$. We may then study the behavior of the recursion (5.2) and make inferences about the behavior of $x^i(t)$.

The above framework covers both specialized processes, in which case we have $L=n$, as well as the case of total overlap where we have $L=1$ and we do not distinguish between components of the estimates. For specialized processes (e.g. example 3) it is easy to see that $y(t) = (x_1^1(t), x_2^2(t),\ldots,x_n^n(t))$.

We now proceed to present some general convergence results. We allow the exogenous measurements z_i^i of each processor, as well as the initialization $x^i(1)$ of the algorithm to be random (with finite variance). We assume that they are all defined on a probability space (Ω,F,P) and we denote by F_t the σ-algebra generated by $\{x^i(1),z_i^i(s); i=1,\ldots,n; s=1,\ldots,t-1\}$. We assume, however, that the sequence of times at which measurements are obtained, computations are performed, the times $\tau_{j,\ell}^i(t)$, as well as the combining coefficients $\beta_{j,\ell}^i(t)$ are deterministic. (In fact, this assumption may be often relaxed). In order to quantify the speed of convergence of $\phi^{ij}(t|s)$ we introduce

$$c(t|s) = \max_{i,j} |\phi^{ij}(t|s) - \phi^j(s)| .$$

By Assumption 5.1 $\lim_{t\to\infty} c(t|s)=0$ and it may be shown that $c(t|s)\leq 1$, $\forall t,s$. Consider the following assumptions:

Assumption 5.2:

$$E[< \frac{\partial f}{\partial x}(x^i(t)), \phi^i(t)z^i_i(t)>|F_t] \leq 0, \qquad \forall t,i, \quad a.s.$$

Assumption 5.3:

a) For some $K_o \geq 0$

$$E[||z^i_i(t)||^2] \leq - K_o E[< \frac{\partial f}{\partial x}(x^i(t)), \phi^i(t)z^i_i(t)>], \quad \forall i,t.$$

b) For some $B \geq 0$, $d \in [0,1)$, $c(t|s) < Bd^{t-s}$, $\forall t \geq s$, $\forall s$.

Assumption 5.2 states that $\phi^i(t)z^i_i(t)$ (which is the "effective update direction" of processor i, see (5.2)) is a descent direction with respect to f. Assumption 5.3a requires that $z^i_i(t)$ is not too large. In particular any noise present in $z^i_i(t)$ can only be "multiplicative-like": its variance must decrease to zero as a stationary point of f is approached. For example, we may have

$$z^i_i(t) = -\left[\frac{\partial f}{\partial x}(x^i(t))\right] (1+w^i(t)),$$

where $w^i(t)$ is scalar white noise. Finally, Assumption 5.3b requires that the processors tend to agree exponentially fast. Effectively, this requires that the time between consecutive measurements of the type $z^i_{j,\ell}$, $i \neq j$, as well as the delays $t-\tau^i_{j,\ell}(t)$ are bounded together with some minor restriction of the coefficients $\beta^i_{j,\ell}(t)$ for those times that a measurement of type $z^i_{j,\ell}$ is obtained.

Letting

$$\alpha_o = \sup_{t,i} \alpha^i(t),$$

we may use Assumptions 5.3a and 5.3b to show that $||x^i(t)-y(t)||$ is of the order of α_o. Using the Lipschitz continuity of $\frac{\partial f}{\partial x}$ it follows that $||\frac{\partial f}{\partial x}(y(t)) - \frac{\partial f}{\partial x}(x^i(t))||$ is also of the order of α_o; then, using Assumption 5.2, it follows that (5.2) corresponds to a descent algorithm up to first order in α_o. Choosing α_o small enough, convergence may be shown by invoking the supermartingale convergence theorem. The above argument can be made rigorous and yields the following proposition (the proof may be found in [35] and in a forthcoming publication):

Proposition 5.1: If Assumptions 5.1, 5.2, 5.3, hold and if α_o is small enough, then:

a) $f(x^i(t))$, $i=1,\ldots,n$, as well as $f(y(t))$ converge, almost surely, and to the same limit.

b) $\lim_{t\to\infty} (x^i(t)-y(t))=0$, $\forall i$, almost surely and in the mean square.

c) $\sum_{t=1}^{\infty} \sum_{i=1}^{n} \alpha^i(t)E[< \frac{\partial f}{\partial x} (x^i(t)), \phi^i(t)z_i^i(t)>|F_t]>-\infty$,   (5.3)

almost surely. The expectation of the above expression is also finite.

A related class of algorithms arises if the noise in $z_i^i(t)$ is allowed to be additive, e.g.

$$z_i^i(t) = \frac{\partial f}{\partial x} (x^i(t)) + w^i(t),$$

where $w^i(t)$ is zero-mean white. In such a case, an algorithm may be convergent only if $\lim_{t\to\infty} \alpha^i(t)=0$. In fact, $\alpha^i(t)=1/t_i$, where t_i is the number of times up to time t that a new measurement was received at i, is the most convenient choice, and this is what we assume here. However, this choice of stepsize implies that the algorithm becomes progressively slower, as $t\to\infty$. We may therefore allow the agreement process to become progressively slower as well, and still retain convergence. In physical terms, the time between consecutive measurements $z_{j,\ell}^i (i\neq j)$ may increase to infinity, as $t\to\infty$. In mathematical terms:

<u>Assumption 5.4</u>: a) For some K_0, K_1, $K_2 \geq 0$,

$$E[||z_i^i(t)||^2] \leq - K_0E[< \frac{\partial f}{\partial x} (x^i(t)), \phi^i(t)z_i^i(t)>] +$$

$$+ K_1E[f(x^i(t))] + K_2$$

b)   For some $B>0$, $\delta\in(0,1]$, $d\in[0,1)$

$$c(t|s)<Bd^{t^\delta-s^\delta}, \quad \forall t\geq s, \forall s.$$

We then have [35]:

<u>Proposition 5.2</u>: Let $\alpha^i(t)=1/t_i$, where t_i is the number of times up to time t that a new measurement was received at i, and assume that for some $\varepsilon>0$, $t_i\geq\varepsilon.t$ for all i,t. Assume also that Assumptions 5.1, 5.2, 5.4 hold. Then the conclusions (a), (b), (c) of Proposition 5.1 remain valid.

Proposition 5.1, 5.2 do not prove yet convergence to the optimum (suppose, for example, that $z_i^i(t)\equiv0$, $\forall i,t$). However, (5.3) may be exploited to yield optimality under a few additional assumptions:

<u>Corollary</u>: Let the assumptions of either Proposition 5.1 or 5.2 hold. Let T^i be the set of times that processor i obtains a measurement of type z_i^i. Suppose that there exists some $B>0$ and, for each i, a sequence $\{t_k^i\}$ of distinct elements of T^i such that

$$\max_{i,j} |t_k^i - t_k^j| \le B, \tag{5.4}$$

$$\sum_{k=1}^{\infty} \min_i \{\alpha^i(t_k^i)\} = \infty .$$

Finally, assume that there exist uniformly continuous functions: $g^i : x \to [0,\infty)$ satisfying

a) $\displaystyle\lim_{|x| \to \infty} \inf \sum_{i=1}^{n} g^i(x) > 0$

b) $E[< \frac{\partial J}{\partial x} (x^i(t)), \phi^i(t)z_i^i(t) > | F_t] \le - g^i(x^i(t)), \quad \forall t \in T^i, \forall i, \text{ almost surely.}$

c) $\displaystyle\sum_{i=1}^{n} g^i(x^*) = 0 \Rightarrow x^* \in X^* \triangleq \{x \in X | f(x^*) = \inf_x f(x)\}$

Then, $\displaystyle\lim_{t \to \infty} f(x^i(t)) = \inf_x f(x), \quad \forall i, \text{ almost surely.}$

Example 3: (continued): It follows from the above results that the distributed deterministic gradient algorithm applied to a convex function converges provided that a) The stepsize α is small enough, b) Assumption 5.3(b) holds and c) The processors update, using (3.7), regularly enough, i.e. condition (5.4) is satisfied. Similarly, convergence for the distributed stochastic gradient algorithm follows if we choose a stepsize $\alpha^i(t) = 1/t_i$, if Assumption 5.4 and condition (5.4) hold.

Example 4: (continued) Similarly with the previous example, convergence to stationary points of f may be shown, provided that α_i is not too large, that the delays $t - \tau_j^{im}(t)$ are not too large and that the processors do not update too irregularly. It should be pointed out that a more refined set of sufficient conditions for convergence may be obtained, which links the "coupling constants" $K_{j,m}^i$ with bounds on the delays $t - \tau_j^{im}(t)$ [35]. These conditions effectively quantify the notion that the time between consecutive communications and communication delays between decision makers should be inversely proportional to the strength of coupling between their respective divisions.

Example 7: (continued) Several common algorithms for identification of a moving average process satisfy the conditional descent Assumption 5.2. (e.g. the Least Mean Squares algorithm, or its normalized version-NLMS). Consequently, Proposition 5.2 may be invoked. Using part (c) of the Proposition, assuming that the input is sufficiently rich and that enough messages are exchanged, it follows that the distributed algorithm will correctly identify the system. A detailed analysis is given in [35].

A similar approach may be taken to analyze distributed stochastic algorithms in which the noises are correlated and Assumption 5.2 fails to hold. Very few global convergence results are available even for centralized such algorithms [34,36] and it is an open question whether some distributed versions of them also converge. However, as in the centralized case one may associate an ordinary differential equation with such an algorithm as in [37,38], and prove local convergence subject to an assumption that the algorithm returns infinitely often to a bounded region (see [35]). Such results may be used, for example, to demonstrate local convergence of a distributed extended least squares (ELS) algorithm, applied to the ARMAX identification problem in Example 7.

6. Convergence of Distributed Processes with Bayesian Updates

In Sections 4 and 5 we considered distributed processes in which a solution is being successively approximated, while the structure of the updates is restricted to be of a special type. In this section we take a different approach and we assume that the estimate computed by any processor at any given time is such that it minimizes the conditional expectation of a cost function, given the information available to him at that time. Moreover, all processors "know" the structure of the cost function and the underlying statistics, and their performance is only limited by the availability of posterior information. Whenever a processor receives a measurement z_j^i (possibly containing an earlier estimate of another processor) his information changes and a new estimate may be computed.

Formally, let $X=R^m$ be the feasible set, (Ω,F,P) a probability space and $f: X \times \Omega \to [0,\infty)$ a random cost function which is strongly convex in x for each $\omega \in \Omega$. Let $I^i(t)$ denote the information of processor i at time t, which generates a σ-algebra $F_t^i \subset F$. At any time that the information of processor i changes, he updates his estimate according to

$$x^i(t+1) = \arg\min_{x \in X} E[f(x,\omega)|F_t^i] \tag{6.1}$$

Assuming that f is jointly measurable, this defines an almost surely unique, F_t^i - measurable random variable [39].

The information $I^i(t)$ of processor i may change in one of the following ways:

a) New exogenous measurements $z_i^i(t)$ are obtained, so that $I^i(t) = (I^i(t-1), z_i^i(t))$.

b) Measurements $z_j^i(t)$ with the value of an earlier estimate of processor i are obtained; that is,

$$z_j^i(t) = x^j(\tau_j^i(t)); \quad \tau_j^i(t) \leq t$$
$$I^i(t) = (I^i(t-1), z_j^i(t)) \tag{6.2}$$

c) Some information in $I^i(t-1)$ may be "forgotten"; that is, $I^i(t) \subset I^i(t-1)$ (or $F_t^i \subset F_{t-1}^i$).

The times at which measurements are obtained as well as the delays are either deterministic or random; if they are random, their statistics are described by (Ω,F,P) and these statistics are known by all processors.

Case 1: Increasing Information. We start by assuming that information is never forgotten, i.e. $F_{t+1}^i \supset F_t^i$, $\forall i,t$. Let $f(x,\omega) = ||x-x^*(\omega)||^2$, where $x^*:\Omega \to R^m$ is an unknown random vector to be estimated. Then,

$$x^i(t+1) = E[x^*(\omega)|F_t^i]$$

and by the martingale convergence theorem, $x^i(t)$ converges almost surely to a random variable y^i. Moreover it has been shown that if "enough" measurements of type (6.2) are obtained by each processor, then $y^i = y^j$, $\forall i,j$, almost surely [30,41]. If f is not quadratic but strongly convex, the same results are obtained except that convergence holds in the sense of probability and in the $L^2(\Omega, F, \mu)$ sense, where μ is a measure equivalent to P, determined by the function f [39]. However, this scheme is not, strictly speaking, iterative, since $I^i(t)$ increases, and unbounded memory is required.

Case 2: Iterative Schemes

The above scheme can be made iterative if we allow processors to forget their past information. For example, let

$$
I^i(t) = \begin{cases} \{x^i(t), z^i_j(t)\}, & \text{if a measurement } z^i_j(t) \text{ is obtained at time t} \\ \{x^i(t)\}, & \text{otherwise .} \end{cases}
$$

Let $z^i_j(t) = x^j(\tau^i_j(t))$, $i \neq j$, $\tau^i_j(t) \leq t$. Assuming that "enough" measurements of this type are obtained by each processor, asymptotic agreement may be still obtained, as for Case 1 [39]. It has been also shown that $x^i(t+1) - x^i(t)$ converges to zero, for each i, but it is not known whether $x^i(t)$ is guaranteed to converge or not.

Even though this case corresponds to an iterative algorithm, it may be very hard to implement: The computation of the minimum in (6.1) may be intractable. Also, even if the processors asymptotically converge and agree, there are no guarantees in general about the quality of the final estimate. There is one notable exception where these drawbacks disappear, which we discuss below:

Case 3: Distributed Linear Estimation

Let $f(x,\omega) = ||x - x^*(\omega)||^2$, where x^* is a zero-mean Gaussian scalar random variable to be estimated. Suppose that at time zero each processor obtains measurements

$$
z^i_{i,k} = x^* + w^i_k, \quad k = 1, \ldots, m_i, \tag{6.3}
$$

where w^i_k are zero-mean Gaussian noises. We allow the noises of different processors to be correlated to each other. Let $I^i(0) = \{z^i_{i,k} | k = 1, \ldots, m_i\}$. No further measurements of the form (6.3) are obtained after time zero. Subsequently each processor i receives from time to time measurements $z^i_j(t) = z^j(\tau^i_j(t))$, $\tau^i_j(t) \leq t$, of the other processors' estimates and updates according to

$$
x^i(t+1) = E[x^* | I^i(0), z^i_j(t)] .
$$

The timing and delay of these latter measurements is assumed to be deterministic. If we make the assumption that an infinite number of measurements of each type z^i_j is obtained by each processor i, together with an additional assumption that essentially requires that there exists an indirect communication path between every pair of processors then it can be shown that $x^i(t)$ converges in the mean square to the centralized estimate

$$x^* = E[x^* | I^1(0), \ldots, I^n(0)] ,$$

which is the optimal estimate of x^* given the total information of all processors [35], [39].

What is interesting about the above algorithm is that it corresponds to a distributed iterative decomposition algorithm for solving the centralized linear estimation problem. The minimization of the cost criterion over a space of dimension $\sum_{i=1}^{n} m_i$, in general, is substituted by a sequence of minimizations along (m_i+1)-dimensional subspaces.

If the noises w_k^i, w_ℓ^i, $i \neq j$, are independent the algorithm converges after finitely many iterations. In general, the algorithm converges linearly but the rate of convergence depends strongly on the number of processors and the angles between certain subspaces of random variables (essentially on the correlations between w_k^i and w_ℓ^i, $i \neq j$, see [35], [39]).

REFERENCES

[1]   Tannenbaum, A.S., Computer Networks, Prentice Hall, Englewood Cliffs, N.J., 1981.

[2]   Schwartz, M., Computer Communication Network Design and Analysis, Prentice Hall, Englewood Cliffs, N.J., 1977.

[3]   Kleinrock, L., Queuing Systems, Vol. I & II, J. Wiley, N.Y., 1975.

[4]   Schwartz, M., and Stern, T.E., "Routing Techniques Used in Computer Communication Networks," IEEE Trans. on Communications, Vol. COM-28, 1980, pp. 539-552.

[5]   Gallager, R.G., "A Minimum Delay Routing Algorithm Using Distributed Computation," IEEE Trans. on Communication, Vol. COM-25, 1977, pp. 73-85.

[6]   Bertsekas, D.P., "Optimal Routing and Flow Control Methods for Communication Networks," in Analysis and Optimization of Systems (A. Bensoussan and J.L. Lions, eds.), Springer-Verlag, Berlin and N.Y., 1982, pp. 615-643.

[7]   Kung, H.T., "Synchronized and Asynchronous Parallel Algorithms for Multiprocessors," in Algorithms and Complexity, Academic Press, 1976, pp. 153-200.

[8]   Bertsekas, D.P., and Gafni, E.M., "Projected Newton Methods and Optimization of Multicommodity Flows," M.I.T., LIDS Report P-1140, M.I.T., Aug. 1981, IEEE Trans. on Aut. Control, Dec. 1983 .

[9]   McQuillan, J.M., Richer, I., and Rosen, E.C., "The New Routing Algorithm for the ARPANET," IEEE Trans. on Communications, Vol. COM-28, 1980, pp. 711-719.

[10] Chazan, D., and Miranker, W., "Chaotic Relaxation," Linear Algebra and Applications, Vol. 2, 1969, pp. 199-222.

[11] Baudet, G.M., "Asynchronous Iterative Methods for Multiprocessors," Journal of the ACM, Vol. 2, 1978, pp. 226-244.

[12] Bertsekas, D.P., "Distributed Dynamic Programming," IEEE Trans. on Aut. Control, Vol. AC-27, 1982, pp. 610-616.

[13] Bertsekas, D.P., "Distributed Asynchronous Computation of Fixed Points," Math. Programming, Vol. 27, 1983, pp. 107-120.

[14] McQuillan, J., Falk, G., and Richer, I., "A Review of the Development and Performance of the ARPANET Routing Algorithm," IEEE Trans. on Communications, Vol. COM-26, 1968, pp. 1802-1811.

[15] Zangwill, W.I., Nonlinear Programming, Prentice Hall, Englewood Cliffs, N.Y., 1969.

[16] Luenberger, D.G., Introduction to Linear and Nonlinear Programming, Addison-Wesley, Reading, MA, 1973.

[17] Ortega, J.M. and Rheinboldt, W.C., Iterative Solution of Nonlinear Equations in Several Variables, Academic Press, N.Y., 1970.

[18] Polak, E., Computational Methods in Optimization: A Unified Approach, Academic Press, N.Y., 1971.

[19] Poljak, B.T., "Convergence and Convergence Rate of Iterative Stochastic Algorithms," Automation and Remote Control, Vol. 12, 1982, pp. 83-94.

[20] Bertsekas, D.P., Dynamic Programming and Stochastic Control, Academic Press, N.Y., 1976.

[21] Sandell, N.R., Jr., P. Varaiya, M. Athans and M. Safonov, "Survey of De-centralized Control Methods for Large Scale Systems," IEEE Trans. on Aut. Control, Vol. AC-23, No. 2, 1978, pp. 108-128.

[22] Ho, Y.C., "Team Decision Theory and Information Structure," IEEE Proceedings, Vol. 68, No. 6, 1980, pp. 644-654.

[23] Yao, A.C., "Some Complexity Questions Related to Distributed Computing," Proc. of the 11th STOC, 1979, pp. 209-213.

[24] Papadimitriou, C.H. and M. Sipser, "Communication Complexity," Proc. of the 14th STOCH, 1982, pp. 196-200.

[25] Witsenhausen, H.S., "A Counterexample in Stochastic Optimum Control," SIAM J. Control, Vol. 6, No. 1, 1968, pp. 138-147.

[26] Papadimitriou, C.H. and J.N. Tsitsiklis, "On the Complexity of Designing Distributed Protocols," Information and Control, Vol. 53, 1982, pp. 211-218.

[27] Tenney, R.R. and N.R. Sandell, Jr., "Detection with Distributed Sensors," IEEE Trans. on Aerospace and Electronic Systems, Vol. AES-17, No. 4, 1981, pp. 501-509.

[28] Willsky, A.S., M. Bello, D.A. Castanon, B.C. Levy and G. Verghese, "Combining and Updating of Local Estimates and Regional Maps along Sets of One-Dimensional Tracks," IEEE Trans. on Aut. Control, Vol. AC-27, No. 4, 1982, pp. 799-813.

[29] Sanders, C.W., E.C. Tacker, T.D. Linton, R.Y.-S. Ling, "Specific Structures for Large Scale State Estimation Algorithms Having Information Exchange," IEEE Trans. on Aut. Control, Vol. AC-23, No. 2, 1978, pp. 255-261.

[30] Borkar, V. and P. Varaiya, "Asymptotic Agreement in Distributed Estimation," IEEE Trans. on Aut. Control, Vol. AC-27, No. 3, 1982, pp. 650-655.

[31] Arrow, K.J. and L. Hurwicz, "Decentralization and Computation in Resource Allocation," in Essays in Economics and Econometrics, R.W. Pfouts, Ed., Univ. of North Carolina Press, Chapel Hill, NC, 1960, pp. 34-104.

[32] Meerkov, S.M., "Mathematical Theory of Behavior-Individual and Collective Behavior of Retardable Elements," Mathematical Biosciences, Vol. 43, 1979, pp. 41-106.

[33] Astrom, K.J. and P. Eykhoff, "System Identification-A Survey," Automatica, Vol. 7, 1971, pp. 123-162.

[34] Solo, V., "The Convergence of AML," IEEE Trans. on Aut. Control, Vol. AC-24, 1979, pp. 958-963.

[35] Tsitsiklis, J.N., "Problems in Decentralized Decision Making and Computation" Ph.D. Thesis, Dept. of Electrical Engineering and Computer Science, MIT, Cambridge, MA, in preparation.

[36] Nemirovsky, A.S., "On a Procedure for Stochastic Approximation in the Case of Dependent Noise," Engineering Cybernetics, 1981, pp. 1-13.

[37] Ljung, T., "Analysis of Recursive Stochastic Algorithms," IEEE Trans. on Auto. Control, Vol. AC-22, No. 4, 1977, pp. 551-575.

[38] Ljung, L., "On Positive Real Transfer Functions and the Convergence of Some Recursive Schemes," IEEE Trans. on Auto. Control, Vol. AC-22, No. 4, 1977, pp. 539-551.

[39] Tsitsiklis, J.N. and M. Athans, "Convergence and Asymptotic Agreement in
      Distributed Decision Problems," to appear in the IEEE Trans. on Aut. Control,
      1984.

[40] Poljak, B.T. and Y.Z. Tsypkin, "Pseudogradient Adaptation and Training
      Algorithms," Automation and Remote Control, No. 3, 1973, pp. 45-68.

[41] Geanakoplos, J.D. and H.M. Polemarchakis, "We Can't Disagree Forever,"
      Institute for Mathematical Studies in the Social Sciences, Technical
      Report No. 277, Stanford University, Stanford, CA, 1978.

STOCHASTIC INTEGER PROGRAMMING:

THE DISTRIBUTION PROBLEM

A.H.G. Rinnooy Kan

Econometric Institute - Erasmus University Rotterdam - The Netherlands

ABSTRACT

A brief summary is given of recent insights into the distribution problem for structured stochastic integer programming problems, as surveyed
during the Gargnano conference. The application of these results within a
heuristic approach to certain hierarchical planning problems is discussed as
well.

Keywords: Stochastic integer programming, hierarchical planning

1. INTRODUCTION

Stochastic integer programming problems arise whenever an integrality
constraint on some of the decision variables is added to an ordinary stochastic program. The literature on this problem class is sparse, and this is
not surprising. Deterministic integer programming problems have earned a
well deserved reputation for computational intractability and the introduction of stochastic features can hardly be expected to have a positive
influence on the computational burden. The framework of NP-completeness
theory (see, e.g., [Lenstra & Rinnooy Kan 1979]), which provides a formal
counterpart to our intuitions about the inherent difficulties of certain
integer programming problems, extends trivially to many stochastic versions

of these problems in that deterministic parameters often can be seen as a special, degenerate case of random ones. Hence, there is no reason to be optimistic about the prospects for efficient general stochastic integer programming algorithms.

Yet, in a strange fashion, the introduction of stochastic features sometimes seems to make these problems easier rather than harder to solve. The discontinuities and nonconvexities that are so typical for integer programming are smoothed out, and occasionally surprisingly simple analytical expressions emerge for the optimal solution value of stochastic versions of well known classes of integer programming problems. These <u>distribution</u> results are usually of an asymptotic nature, and they lead to the remarkable situation that, as the problem size goes up, increasingly accurate estimates can be made of the optimal solution value of certain integer programs, although the exact solution of a particular large sized instance of that program may well require an inordinate computational effort.

This phenomenon is not yet understood too well and points to a research area of great interest that is currently being explored with great energy. In what follows, an attempt will be made to convey the flavour of some typical recent results. These results are categorized most naturally according to the type of the underlying probabilistic model. We shall in fact distinguish between three types of problems:

- <u>number problems</u>, in which randomness occurs in certain numerical parameters; typically, certain coefficients will be assumed to be i.i.d. random variables (Section 2);
- <u>Euclidean problems</u>, defined with respect to points in the Euclidean plane; typically, these points will be assumed to be uniformly distributed over a certain fixed region (Section 3);
- <u>graph problems</u>, where randomness can be introduced according to one of the two standard models: either a fixed number of edges is distributed randomly among all pairs of vertices or each possible edge is supposed to occur with a fixed probability (Section 4).

For more detailed information, the reader may wish to consult a recent bibliography on probabilistic analysis of algorithms [Karp et al. 1984], in which many additional references can be found.

2. NUMBER PROBLEMS

Two examples will serve to clarify the characteristic features of this class.

MACHINE SCHEDULING (MS): given M identical machines and N jobs with processing times p_1, ..., p_n, distribute the jobs among the machines so as to minimize the makespan of the schedule.

LINEAR ASSIGNMENT (LA): given the cost A_{ij} of assigning man i to job j (i,j = 1,...,n), find the assignment that minimizes total cost.

Natural stochastic versions of both problems are obtained by assuming that the numbers p_j (j=1,...,n) and A_{ij} (i,j=1,...,n) respectively are i.i.d. random variables. We are interested in the distribution of the random variable corresponding to the optimal solution values z_{ms} and z_{la}.

The usual approach to these problems is to search for a simple heuristic solution method of which the error can be shown to grow slower than (a lower bound on) the optimal solution in some probabilistic sense as the problem size n increases.

The scheduling problem provides a good example of this approach. An appropriate heuristic is the LPT rule, in which jobs are assigned to the first available machine in order of nonincreasing p_j. The makespan of the resulting schedule z_{ms}^{lpt}, is compared to the lower bound on z_{ms} given by $\sum_{j=1}^{n} p_j/m$. In [Frenk & Rinnooy Kan 1984], it is shown that if $Ep_j < \infty$,

$$z_{ms}^{lpt} - \sum_{j=1}^{n} p_j/m \to 0 \qquad\qquad \text{(a.s.)} \qquad\qquad (1)$$

and if $Ep_j^2 < \infty$,

$$EZ_{ms}^{lpt} - nEp_j/m \to 0 \qquad\qquad\qquad\qquad (2)$$

(cf. [Loulou 1982]).

The immediate implication of this result is that, under the above conditions, the optimal solution value itself converges to the lower bound value almost surely and in expectation, so that a simple closed form expression provides the solution to the distribution problem for sufficiently large values of n.

Of course this result is only an asymptotic one. However, the speed at which convergence occurs can be estimated for some special distributions. E.g., when the distribution of these p_j is uniform, the speed of convergence of (2) is $1/n$. In fact, by using a different argument it can be shown that for m=2 the optimal solution value converges to the lower bound at exponential speed [Karmarkar et al. 1984].

The distribution result for the scheduling problem confirms our intuition that for n sufficiently large an optimal schedule will be very compact in that the workload Σp_j is divided evenly over the m machines. A similarly simple intuition indicates what the optimal linear assignment value would be expected to be.

It is convenient to think of this problem as defined on a **bipartite graph** with nodes $M_1, \ldots, M_n$, $J_1, \ldots, J_n$ and edges $\{M_i, J_j\}$ with weight A_{ij}. For n sufficiently large, consider the removal of all edges except the three with the smallest weight at every node. The resulting subgraph can be shown to admit of a valid assignment with probability increasing to 1 exponentially fast (see [Walkup 1979]). This implies that the optimal solution value is determined by the small order statistics of the distribution of A_{ij}. Indeed, it is shown in [Frenk & Rinnooy Kan 1984a] that under quite general conditions on the distribution function F, Ez_{la} is asymptotic to $nF^{-1}(1/n)$.

This result implies that if the A_{ij} are uniformly distributed, Ez_{la} is bounded by a constant; this constant was recently shown to be at most 2 [Karp 1984]. For another special case, see [Loulou 1982a].

Generally, the analysis for these number problems strongly depends on the availability of a simple, yet asymptotically optimal heuristic.

Viewed in that light, the **knapsack problem** is a natural candidate for

an analysis of this type; some work on this problem is currently in progress
(see e.g., [Lueker 1980]).

3. EUCLIDEAN PROBLEMS

Some classical combinatorial optimization problems allow a natural
formulation in the Euclidean plane.

TRAVELLING SALESMAN (TS) : given N points in the plane, find the shortest
closed tour.

MEDIAN LOCATION (ML) : given N points in the plane, find k locations among
them such that the total distance between the points and their
nearest location is minimized.

We are interested in the distribution of the optimal solution values
Z_{ts} and Z_{ml} under the assumption that the points are independently uniformly
distributed over a fixed region.

The travelling salesman problem was one of the first problems to be
attacked along those lines. In a classical paper, [Beardwood et al. 1959],
it is shown that

$$Z_{ts}/\sqrt{n} \to B \qquad\qquad (a.s.) \qquad\qquad (3)$$

where B is a constant depending on the shape of the region (for the unit
square, $B \approx 0.765$). The proof technique is complicated; in a simplified
approach in [Karp & Steele 1984] the expected value analysis carried out
for points generated by a Poisson process is transferred to this model
through an Abel-Tauber theorem. It is not hard, however, to appreciate the
result intuitively: a tour through 4n points in a 4x4 square is about 4
times as long as a tour through n points in the unit square; this suggests
that the tour length is doubled when the number of points is multiplied by 4.

A heuristic scheme based on precisely the idea of partitioning the
region into subregions and linking the optimal tours in each of them was
shown to produce asymptotically optimal tours almost surely in [Karp 1977]
(see also [Steele 1981a]). However, the heuristic analysis here presupposes

the optimal value result, unlike the situation in the previous section.

The median location problem is solved by application of a different technique that is a natural one for some Euclidean problems, i.e. by exploiting the relation between discrete problems and their continuous analogues; in this case, continuity extends to both the demand (then assumed to be distributed uniformly over the region) and the possible locations (no longer restricted to the given points).

For this continuous problem it is intuitively clear that an optimal solution will exhibit a great deal of regularity; indeed, it will correspond to the partition of the region in an appropriate number of hexagons with locations appearing in their centers. If n and k go to infinity, with $k=o(n/\log n)$, then the continuous model provides an appropriate approximation and the analysis yields that $\left[\text{Papadimitrion 1980}\right]$

$$\frac{Z_{ml}}{n\sqrt{A/k}} \to \gamma \qquad\qquad \text{(a.s.)} \qquad\qquad (4)$$

where A is the area of the region and γ is the solution value of the continuous problem for one normalized hexagon $\left[\text{Haimovich 1984}\right]$:

$$\gamma = (\frac{2}{3\ \sqrt{3}})^{1/2} (\frac{1}{3} + \frac{1}{4}\ \ell n3).$$

For that case that $k=\alpha n$, the analysis is based on a generalization of the Beardwood et al. result $\left[\text{Steele 1981; Hochbaum \& Steele 1982}\right]$.

Again, a form of partitioning provides the crucial insight into the structure of the optimal solution value for these problems. Unfortunately, nothing is known about the speed of convergence for these results and computational experiments suggest that the performance of the partitioning heuristics for small values of n is rather poor.

4. GRAPH PROBLEMS

Although there is a rich literature on the structure of random graphs

$\left[\text{Erdös \& Spencer } 1974\right]$, many of the results reported there do not bear directly on optimization problems. As a particularly fine example of what is known in the latter context, we consider the following problem.

MAXIMUM CLIQUE (MC): given an undirected graph, G = (V,E) find the complete subgraph with the largest number of vertices.

The deterministic version of this problem is notoriously difficult to solve. Again, however, a natural randomization leads to a distribution problem for which many analytical results have been obtained.

Let us assume that, for given n, each possible edge between two vertices is present with probability p. Then the following results are known about $\underline{Z}_{mc}$, the size of the maximum clique.

$$\frac{Z_{mc}}{\ell n \; n} \rightarrow \frac{2}{\ell n \; 1/p} \qquad \text{(a.s.)} \qquad (5)$$

and if

$$t(n,p) = 2 \, \log_{1/p} n - 2 \, \log_{1/p} \log_{1/p} n + 2 \, \log_{1/p} e/2 + 1 \qquad (6)$$

then

$$\lim_{n \to \infty} \Pr\left\{\left[t(n,p) - \varepsilon\right] \le \underline{Z}_{mc} \le \left[t(n,p) + \varepsilon\right]\right\} = 1 \qquad (7)$$

A remarkable feature of these results is the peakedness of the distribution of $\underline{Z}_{mc}$, as witnessed by (5) and (7). We note also that, contrary to the results presented in Sections 2 and 3, the above analysis does not involve or lead to a heuristic that solves the problem to asymptotic optimality.

A limited literature also exists on optimization problems defined on complete graphs with random edge weights (e.g. $\left[\text{Lueker } 1981\right]$).

5. APPLICATION TO TWO STAGE PROBLEMS

Certain types of distribution results find application in the context of two stage stochastic integer programming problems.Here, the first stage would typically correspond to the acquisition of a certain amount of resource subject to uncertainty with respect to its actual usage, and the second stage would correspond to the actual allocation of the resource, given precise information about the demand for its use. Both the first and second stage may involve integrality constraints on the decision variables; this is particularly true for the second stage, where the resource allocation problem will often be of a combinatorial nature.

Now, if appropriate information about the second stage distribution problem is available, a natural heuristic suggests itself to solve the two stage problem. If, for sufficiently large problem size, the optimal solution value can be written as a simple function of some problem parameters including the amount of resource involved, then a first stage heuristic would be to choose the amount of resource so as to minimize first stage acquisition cost plus the asymptotic second stage allocation cost, and a second stage heuristic would be to allocate the resource using any asymptotically optimal heuristic.

The simplest example of such an approach arises in a hierarchical scheduling problem, where in the first stage the number m of machines to be acquired at cost c each has to be determined, subject to uncertainty about the processing times $p_1, \ldots, p_n$ of the n jobs to be executed on the machines in the second stage. If the allocation cost is assumed to be proportional to the makespan of the schedule, then we may use the result from Section 2, approximate the expected makespan by $nE\underline{p}_j/m$ and chose m to minimize $cm + nE\underline{p}_j/m$. This first stage stochastic programming heuristic combined with, say, the LPT rule in the second stage has strong properties of asymptotic optimality [Dempster et al. 1983]. Related models are discussed in [Marchetti et al. 1983, Frenk et al. 1982].

The natural way in which distribution results finds application in this hierarchical planning context provides additional motivation for continuing research in this area.

6. CONCLUDING REMARKS

The selection of results obtained for special integer programming problems is not intended to disguise that little or no progress has been made on algorithms to solve the general stochastic integer programming problem. If only as a theoretical challenge, this problem deserves serious consideration. The positive side effects of stochasticity that were so noticeable in all special cases may carry over to this general problem as well.

REFERENCES

J. Beardwood, J.H. Halton & J.M. Hammersley (1959), "The shortest path through many points", Proc. Cambridge Philos. Soc. 55, 299-327.

M.A.H. Dempster, M.L. Fisher, L. Jansen, B.J. Lageweg, J.K. Lenstra & A.H.G. Rinnooy Kan (1983) "Analysis of heuristics for stochastic programming: results for hierarchical scheduling problems", Math. Op. Res. 8, 525-538.

P. Erdős & J. Spencer (1974), Probabilistic Methods in Combinatorics, Academic Press.

J.B.G. Frenk & A.H.G. Rinnooy Kan (1984), "The asymptotic optimality of the LPT rule", Technical report, Erasmus University Rotterdam.

J.B.G. Frenk & A.G.H. Rinnooy Kan (1984a), "Order statistics and the linear assignment problem", Technical Report, Erasmus University Rotterdam.

J.B.G. Frenk, A.H.G. Rinnooy Kan & L. Stougie (1984), "A hierarchical scheduling model with a well solvable second stage", Annals of O.R. 1 (to appear).

M. Haimovich (1984), Ph. D. Thesis, M.I.T.

D.S. Hochbaum & J.M. Steele (1982), "Steinhaus" geometric location problem for random samples in the plane", Adv. in Appl. Prob. 14, 59-67.

N. Karmarkar, R.M. Karp, G.S. Lueker & A. Odlyzko (1984), "The probabiility distribution of the optimal value of a partitioning problem", Manuscript, Bell Laboratories.

R.M. Karp (1977), "Probabilistic analysis of partitioning algorithms for the traveling-salesman problem in the plane", Math. Op. Res. 2, 209-224.

R.M. Karp (1984), "An upper bound on the expected cost of an optimal assignment". Manuscript, Computer Science Division, University of California, Berkeley.

R.M. Karp, J.K. Lenstra, C. Mc Diarmid & A.H.G. Rinnoy Kan (1984), "Probabilistic analysis of algorithms: an annotated bibliography" in: M. Oh. Eigeartaigh et al. (eds.) (1985), Combinatorial Optimization: Annotated Bibliographies, Wiley.

R.M. Karp & J.M. Steele (1984), "Probabilistic analysis of the traveling salesman problem" in E. Lawler et al. (eds.) (1985), The Traveling Salesman Problem, Wiley.

J.K. Lenstra & A.H.G. Rinnoy Kan (1979), "Computational complexity of discrete optimization problems", Ann. Discr. Math. 5, 287-326.

R. Loulou (1982), "Probabilistic behaviour of optimal bin packing solutions", Technical Report, Mc Gill University, Montreal.

R. Loulou (1982a), "Average behaviour of heuristic and optimal solutions to the maximization assignment problem", Technical Report, Mc Gill University, Montreal.

G.S. Lueker (1980), "On the average difference between the solutions to linear and integer knapsack problems" in: R.L. Disney & T.J. Ott (eds.) (1982), Applied Probability - Computer Science: The Interface, Volume I, Birkhäuser Boston.

G.S. Lueker (1981), "Optimization problems on graphs with independent random edge weights", SIAM J. Comp. 10, 338-351.

A. Marchetti Spaccamela, A.H.G. Rinnooy Kan & L. Stougie (1984), "Hierarchical vehicle routing problems", Networks (to appear).

D.W. Matula (1976), "The largest clique size in a random graph, Technical Report, Southern Methodist University, Dallas.

C.H. Papadimitriou (1980), "Worst case and probabilistic analysis of a geometric location problem", Technical Report, M.I.T.

J.M. Steele (1981a), "Complete convergence of short paths and Karp's algorithm for the TSP", Math. Op. Res. 6, 374-378.

J.M. Steele (1981), "Subadditive Euclidean functionals and nonlinear growth in geometric probability", Ann. Probability 9, 365-376.

D.W. Walkup (1979), "On the expected value of a random assignment problem", SIAM J. Comp. 8, 440-442.

THE DUALITY BETWEEN EXPECTED UTILITY AND PENALTY

IN STOCHASTIC LINEAR PROGRAMMING

Aharon Ben-Tal* and Marc Teboulle
Faculty of Industrial Engineering & Management
Technion-Israel Institute of Technology, Haifa

ABSTRACT: We study the dual problem corresponding to a linear program in which the stochastic objective function is replaced by its expected utility, and discuss its relevance as a penalty method to a stochastically constrained dual linear program.

1. INTRODUCTION

Consider a linear programming problem with a stochastic objective function:

$$(\text{SO-P}) \qquad \sup\{c^t x : Ax \le b , x \ge 0\}$$

where $x \in R^n$ is the decision vector, A is a given fixed matrix of size $m \times n$, b is a given fixed vector in R^m, and c is a random vector with a known distribution function $F_c(.)$, and mean $E(c) = \mu \in R^n$.

Problem (SO-P) (and likewise its dual (DC-D) below) is of course meaningless, unless a decision-theoretic approach is associated with it. One of the fundamental approaches to solve (SO-P) is the well known <u>expected utility principle</u> (see for example [9], [5]) which reduces problem (SO-P) to the deterministic nonlinear programming problem:

$$(\text{EU-P}) \qquad \text{Sup}\{Eu(c^t x) : Ax \le b , x \ge 0\}$$

where E denotes the mathematical expectation with respect to the random vector c, and u is the decision maker utility function, which is strictly increasing and strictly concave. The latter property reflects a risk-aversion attitude. (see e.g. the papers of Arrow [1] and Pratt [6]). Under these conditions problem (EU-P) becomes a concave nonlinear programming problem.

The purpose of this paper is to investigate the dual problem of (EU-P) and to discuss its relations to the dual linear program of (SO-P):

$$(\text{SC-D}) \qquad \min\{b^t y : A^t y \ge c , y \ge 0\} .$$

which is a linear program with <u>stochastic constraints</u>. In fact we find it appropriate to deal, instead of problem (EU-P), with an equivalent problem

---

* Research supported by NSF Grant No. ECS-821408] and Technion VPR-Fund.

(CE-P) $\sup\{u^{-1}Eu(c^tx) : Ax \le b , x \ge 0\}$

where u^{-1} is the inverse of u . Note that for a random variable T

$$C(T) \triangleq u^{-1}Eu(T)$$

is the so-called <u>certainty equivalent</u> of T: it is the sure amount which leaves the
decision maker indifferent to a gamble yielding T, since

$$u(C(T) = Eu(T) .$$

There are few reasons to prefer the formulation (CE-P) to (EU-P):

(i)   In the deterministic case (c being a degenerate random vector) problem (CE-P)
reduces exactly to the original problem (SO-P), and so the linear structure
of it (and its dual) are recovered.

(ii) According to the Von-Neuman-Morgenstern theory [8], the utility function is
unique up to a monotone increasing affine transformation. If the (EU-P)
formulation is used, the dual problem will depend on the particular choice
of u . This is not the case with the (CE-P) formulation since C(T) is
<u>invariant to affine transformation in u.</u>

There is however some difficulty associated with the objective function

$$V(x) = C(c^tx) = u^{-1}Eu(c^tx)$$

of (CE-P); since u^{-1} is convex, it is not guaranteed that V(x) is a concave function. We show however, in the next section, that for a large class of utility functions, including the important class of HARA utilities (see [2], [3], [5], [9]) V(x)
is concave regardless of the distribution of the random vector c .
Properties of the dual problem of (CE-P) are derived in section 3. It is shown there
that the dual of (CE-P) is given by

(CE-D) $\text{Inf}\{y^tb+P_u(y)\}$.
           $y \ge 0$

The first term y^tb is just the objective function of (SC-D). We show that the
second term:

$$P_u(y) = \sup\{V(x)-y^tAx\} .$$
$$ x \ge 0$$

penalizes solution of (SC-P) which are not "feasible in the mean". In section 4 we
introduce our approach to treat linear program with stochastic constraints (SC-D).

The approach is based on the properties of P_u derived in the previous section. For computational conveniences we advocate there the use of additive penalty where each stochastic constraints is individually penalized. We also show how to incorporate in the penalty term the objective attitude towards risk of the decision maker. Useful mean-variance approximations for P_u are given in the last section.

2. CONCAVITY OF THE CERTAINTY EQUIVALENCE FUNCTIONAL

We assume now and henceforth in this paper that

(i) The random vector $c = (c_1, c_2, \ldots, c_n)$ is non-degenerate, i.e. $\forall x \neq 0$ $c^t x$ is not a degenerate univariate random variable.

(ii) $E(c^t x) < +\infty \ \forall x \in R^n$

We denote by $[\underline{c}_k, \overline{c}_k]$ the support of the random variable, by μ_k its mean and by σ_k its standard deviation. Let U be the class of twice differentiable strictly concave utility function with $u' > 0$ and $u'' < 0$.
For a given $u \in U$, the inverse function u^{-1} exists, and it is a strictly monotone increasing function. Thus problem (EU-P) can be transformed into an equivelent one:

$$(CE-P) \qquad \sup\{V(x) \overset{\Delta}{=} u^{-1}Eu(c^T x) \ : \ Ax \leq b \ , \ x \geq 0\}$$

Problem (CE-P) is called the <u>certainty equivalent problem</u>.
Note that $V(x)$ is a convex increasing transformation (u^{-1}) of a concave function $Eu(c^t x)$, thus in general $V(\cdot)$ is not necessarily a concave function. To be able to use the powerful duality theory of convex programming, it is then of major importance to study conditions under which $V(x)$ is concave.
The next result furnish a complete answer to this question. Surprisingly, the concavity of V (for arbitrary distribution of the random vector c) is fully characterized in term of the so-called <u>Arrow-Pratt local risk aversion indicator</u>:

$$r(t) \overset{\Delta}{=} - \frac{u''(t)}{u'(t)} \qquad\qquad (2.1)$$

<u>Theorem 2.1.</u> Let u be a utility function in U. Then the function:

$$V(x) = u^{-1}Eu(c^T x)$$

is concave <u>for any random vector c</u> if and only if $\dfrac{1}{r(t)}$ is concave.

<u>Proof:</u> (The details will appear elsewhere in a future paper and thus are omitted).
[outline]: (i) The concavity of V is established by showing that it satisfies the gradient inequality (see [7]). Denoting by $\varphi(t) \overset{\Delta}{=} u^{-1}(t)$, and using the fact that $\varphi' > 0$, the gradient inequality here is:

$$\forall x,y \in R^n \;\; ; \;\; \frac{\varphi Eu(c^T y) - \varphi Eu(c^T x)}{\varphi'(Eu(c^T x))} \leq E[\frac{c^T(y-x)}{\varphi'[u(c^T x)]}] \tag{2.2}$$

(ii) Given $u \in U$ we define $h : R^2 \to R$ by

$$h(t_1,t_2) \overset{\Delta}{=} \frac{\varphi(t_1) - \varphi(t_2)}{\varphi'(t_2)}$$

and show that h is a convex function if and only if $\frac{1}{r(t)}$ is concave.

(iii) Applying Jensen inequality to $h(t_1,t_2)$ with $t_1 = u(c^t y)$ and $t_2 = u(c^t x)$ we get the inequality (2.2).

□

Remark 2.1   The two parameter class of utility function called hyperbolic absolute risk aversion (HARA) defined by $r(t) = \frac{1}{at+b}$, which is widely used in economics ([2], [5], [9] satisfies trivially the condition that $\frac{1}{r(t)}$ is concave.

The HARA family consists of the following utilities (defined for $t > -b/a$)

$$u(t) = \begin{cases} -e^{-t/b} & \text{if} \quad a = 0, \; b \neq 0 \\ \log(b+t) & \text{if} \quad a = 1 \\ (at+b)^{(a-1)/a} & \text{if} \quad a \neq 0, \; a \neq 1 \end{cases}$$

A utility function satisfying the condition that $1/r(t)$ is concave will be termed a CRT-utility (concave risk tolerance).

3. THE DUAL PROBLEM: AN INDUCED u-PENALTY

The dual problem of

$$\text{(CE-P)} \qquad \sup\{V(x) \overset{\Delta}{=} u^{-1} Eu(c^t x) \; : \; Ax \leq b \; , \; x \geq 0\}$$

is derived via Lagrangian duality. We assume that u is a CRT-utility so (CE-P) is a concave program. The Lagrangian for problem (CE-P) is the function $L:R_+^n \times R_+^m \to R$ with values:

$$L(x,y) = V(x) + y^t(b-Ax) \tag{3.1}$$

The dual objective function is defined by:

$$D(y) = \sup_{x \geq 0}\{V(x) + y^t(b-Ax)\} \tag{3.2}$$

and thus the dual problem for (CE-P) is given by:

$$\text{(CE-D)} \qquad \inf_{y \geq 0} y^t b + \sup_{x \geq 0} \{ V(x) - y^t A x \} \} \qquad\qquad (3.3)$$

Let

$$P_u(y) \overset{\Delta}{=} \sup_{x \geq 0} \{ V(x) - y^t A x \} \qquad\qquad (3.4)$$

Then problem (CE-D) becomes:

$$\text{(CE-D)} \qquad \inf_{y \geq 0} \{ y^t b + P_u(y) \}$$

The first term, $y^t b$ is just the objective function of the dual problem of (SO-P):

$$\text{(SC-D)} \qquad \inf_{y > 0} \{ b^t y \; : \; A^t y \geq c \}$$

We show below that the second term, $P_u(y)$ plays the role of a penalty function for stochastic constraints in (SC-D), P_u will be called accordingly u-penalty.

<u>Theorem 3.1</u> The u-penalty function (3.5) is convex and satisfies:

(i) $P_u(y) = 0$ if $A^t y \geq \mu$ $\qquad\qquad (3.7)$

(ii) $P_u(y) > 0$ if $A^t y \not\geq \mu$ $\qquad\qquad (3.8)$

<u>Proof</u>: As a pointwise supremum of affine functions, $P_u(y)$ is convex

(i) Let $Q(y;x) \overset{\Delta}{=} V(x) - y^t A x$ then

$$P_u(y) = \sup_{x \geq 0} Q(y;x) \geq Q(y;0) = 0 \qquad\qquad (3.9)$$

since $u \in U$, by Jensen inequality $v(x) = u^{-1} Eu(c^t x) \leq \mu^t x$, where equality holds only for $x = 0$, and so

$$P_u(y) = \sup_{x \geq 0} Q(y;x) \leq \sup_{x \geq 0} x^t (\mu - A^t y) \qquad\qquad (3.10)$$

Now if $A^t y \geq \mu$, the inequality (3.10) shows that $P_u(y) \leq 0$ which together with (3.9) proves (3.7).

(ii)   Assume that for some $k(k=1,\ldots,n)$, $A^t_k y < \mu_k$

where A^t_k is the k-th row of A.

Let $Q_k(y;x_k) \triangleq Q(y;0,0,\ldots,x_k,\ldots 0) = V(x_k) - x_k A^T_k y$

Then we have $Q_k(y;0) = 0$ and $\dfrac{d}{dx_k} Q_k(y;0) = \mu_k - A^T_k y > 0$.

Hence there exists $\hat{x}_k > 0$ such that:

$$Q_k(y;\hat{x}_k) > Q_k(y;0) = 0 \tag{3.11}$$

Noting that $P_u(y) = \sup_{R^n \ni x_k \geq 0} Q_k(y;x_k)$ $\tag{3.12}$

and using (3.11), it follows that $P_u(y) > 0$.   □

The theorem demonstrates that $P_u(y)$ penalizes solution of (SC-D) which are not feasible in the mean.

We say that y_1 is less feasible than y_2 for the k-th constraint (in the mean) if:

$$A^t_k y_2 \leq A^t_k y_1 \tag{3.13}$$

and we write it $y_1 \succ y_2$.

The next result shows other desirable properties of the u-penalty. For the k-th constraint we define:

$$P^k_u(y) = \sup_{x_k \geq 0}\{V(x_k) - x_k A^t_k y\} \tag{3.14}$$

Theorem 3.2 The u-penalty function satisfies

(i)   if $y_1 \succ y_2$ then $P^k_u(y_1) > P^k_u(y_2)$ $\tag{3.15}$

(ii) if for some $k:A^t_k y < c_k$ then $P_u(y) = \infty$ $\tag{3.16}$

Proof:

(i)   For $y_1 > y_2$ we have by (3.13):

$$V(x_k) - x_k A'_k y_1 > V(x_k) - x_k A'_k y_2$$

and then using (3.14) the inequality (3.15) holds.

(ii) Let $A^T_k y < c_k$. By (3.12) we have:

$$P_u(y) = \sup_{x_k \geq 0}\{V(x_k) - x_k A_k^t y\} \tag{3.17}$$

since u is increasing and $c_k \geq \underline{c}_k$, (3.17) implies:

$$P_u(y) \geq \sup_{x_k > 0}\{u^{-1}Eu(\underline{c}_k x_k) - x_k A_k^t y\} = \sup_{x_k \geq 0}\{(\underline{c}_k - A_k^t y)x_k\} \tag{3.18}$$

and hence (3.16) follows. □

Theorem 3.2 demonstrates that the u-penalty is <u>monotone</u> in the sense defined in (3.13).

The second property (3.16) shows that the u-penalty has the desirable property to exclude solutions which are not feasible for (SC-D) in probability 1, since for those $P_u(y) = \infty$.

4. LINEAR PROGRAMMING WITH STOCHASTIC CONSTRAINTS: A GENERAL APPROACH

In this section we use the properties of the u-penalty studied previously to outline a general approach for linear programming problem with stochastic constraints:

$$\inf \quad y^t b$$

(SC) $\quad A_k^t y \geq c_k \qquad k = 1,\ldots,n$

$$y \geq 0$$

Suppose that the decision-maker accepts a solution y as being "feasible" for the constraint $A_k^t y \geq c_k$ if $A^t y$ is "large enough", and rejects solutions y for which $A^t y$ is "too small", i.e. he can choose two positive numbers k_1, k_2 such that:

y is feasible if $A_k^t y \geq \mu_k + k_1 \sigma_k$ $^{(*)}$

y is infeasible if $A_k^t y < \mu_k - k_2 \sigma_k$

all other solutions, i.e. y's such that:

$$\mu_k - k_2 \sigma \leq A_k^t y < \mu_k + k_1 \sigma_k$$

are "semi-feasible".

$^{(*)}$ If "Feasibility" is modeled by chance constraint [4], i.e. $\Pr(A_k^t y \geq c_k) \geq \alpha_k$, then $k_1 = F_{z_k}^{-1}(\alpha_k)$ where F_{z_k} is the distribution function of the random variable $z_k = \dfrac{c_k - \mu_k}{\sigma_k}$, and $\alpha_k \in [0,1]$ are prescribed probability levels.

We will now build a penalty function $P_u^k(y)$ for the k-th constraint of (SC) which will reflect the above subjective attitude of the decision-maker.

Let u be a CRT-utility. For fixed k define:

$$\alpha_k = (k_1 + k_2) \cdot \frac{\sigma_k}{\mu_k - \underline{c}_k} \tag{4.1}$$

$$\beta_k = \mu_k - (k_1 \underline{c}_k + k_2 \mu_k) \cdot \frac{\sigma_k}{\mu_k - \underline{c}_k} \tag{4.2}$$

and let c_k^* be the random variable:

$$c_k^* = \alpha_k c_k + \beta_k$$

The penalty function is given by:

$$P_u^k(y) = \sup_{x_k \geq 0} \{u^{-1} E u(c_k^* x_k) - x_k A_k^t y\} \tag{4.3}$$

The coefficients α_k, β_k were chosen so that:

$$E(c_k^*) = \mu_k + k_1 \sigma_k$$

$$\underline{c}_k^* = \mu_k - k_2 \sigma_k$$

hence by Theorem 3.1 and Theorem 3.2 (ii) we have:

$$P_u^k(y) = \begin{cases} 0 & \text{if } A_k^t y \geq \mu_k + k_1 \sigma_k & (4.4) \\ \\ \infty & \text{if } A_k^t y < \mu_k - k_2 \sigma_k & (4.5) \\ \\ \text{positive} & \text{otherwise} & (4.6) \end{cases}$$

Thus semi-feasible solutions cause a positive, but finite penalty. Moreover, by Theorem 3.2 (i), the more they violate feasibility, the more they are penalized. Thus, the approach we advocate for treating the stochastic linear program (SC) is to use the surrogate deterministic program (P):

$$(P) \quad \inf_{y \geq 0} \{y^t b + \sum_{k=1}^n P_u^k(y)\} \tag{4.7}$$

In comparison with problem (CE-D) in (3.5), we use here an additive penalty (sum of penalties for individual constraints) rather than a joint constraints penalty. The additive form is clearly advantageous from the computational viewpoint. It is applicable whenever the decision maker <u>can</u> treat the constraints individually. However there is one choice of the utility function under which the joint constraints penalty <u>is</u> additive:

<u>Theorem 4.1.</u> Let u be an exponential utility function:

$$u(t) = a - be^{-t/p} \qquad (p>0,\ b>0,\ a \in R)$$

If the random variables $c_1, c_2, \ldots, c_n$ are independent then:

$$P_u(y) = \sum_{k=1}^{n} P_u^k(y)$$

where here: $P_u^k(y) = p \cdot \sup_{x_k \geq 0} \{-\log E(e^{-c_k x_k}) - x_k A_k^t y\}$

<u>Proof</u>: The result follows immediately from the fact that in the case of exponential utility the certainty equivalent, in terms of which P_u is defined, is additive. (See, [2], Theorem 4).   □

We close this section by a simple illustrative example. Consider the one dimensional inventory problem:

(SC) $\min\{hy : y \geq d,\ y \geq 0\}$

where h is the unit holding cost and d is the demand. Assume that $d \sim \exp(\lambda)$ with mean $\mu = \frac{1}{\lambda}$ ($\lambda > 0$). Let $u(t) = 1 - e^{-t/p}$ ($p>0$); i.e. the risk-aversion indicator is $r(t) = \frac{1}{p}$. Then by (4.3) the penalty function is:

$$P_u(y) = \sup_{x>0} \{-p \log \frac{\lambda}{\lambda+x} - (\frac{y-\beta}{\alpha})\}$$

where here, by (4.1) - (4.2): $\alpha = k_1+k_2$; $\beta = \mu(1-k_2)$.
By simple calculus we obtain:

$$P_u(y) = \begin{cases} 0 & \text{if } y \geq (1+k_1)\mu \\ p\{\log \mu \frac{k_1 + k_2}{y+\mu(k_2-1)} + \frac{y-\mu(k_2+1)}{\mu(k_1 + k_2)}\} & \text{if } \mu(1-k_2) \leq y < \mu(1+k_2) \\ +\infty & \text{if } y < \mu(1-k_2) \end{cases}$$

Substituting P_u in (4.7) and solving problem (P) we obtain

$$y^* = \mu(1-k_2) + \mu \frac{k_1 + k_2}{1 + \mu(k_1+k_2)\frac{h}{p}}$$

The optimal inventory y^* is a monotone decreasing function of both the holding cost h and the risk-aversion $1/p$. If either $h = 0$ or $1/p \to 0$ (risk neutrality) then y^* equals to the lowest value $\mu(1-k_2)$ while if $h \to \infty$ or $1/p \to \infty$ (extreme risk aversion) y^* equals to the highest value $\mu(1+k_1)$.

5. MEAN VARIANCE APPROXIMATIONS

In this section we derive a quadratic approximation for the u-penalty function $P_u(y)$.
Denote the mean vector of c by μ, and by V its variance-covariance matrix (positive definite).
It can be shown by direct differentiation of the certainty equivalent functional

$$V(x) \overset{\Delta}{=} u^{-1}Eu(c^tx)$$

that:

$$V(0) = 0$$

$$\nabla V(0) = \mu$$

$$\nabla^2 V(0) = -r_0 V \quad \text{with} \quad r_0 \overset{\Delta}{=} -\frac{u''(0)}{u'(0)} \geq 0$$

Using this in (3.4) we obtain:

Proposition 5.1. A second order approximation of $P_u(y)$ is:

$$\hat{P}_u(y) = \sup_{x \geq 0}\{x^t(A^ty-\mu) - \frac{1}{2} r_0 x^t Vx\} \tag{5.1}$$

□

Thus the approximate u-penalty $\hat{P}_u(y)$ is given in term of a concave quadratic program. A direct compution shows that the approximation is exact for the case of exponential utility and c being jointly normal.
For additive penalties, even a more explicit approximation is possible:

Proposition 5.2. A second order approximation $\hat{P}_u(y)$ of $P_u(y) = \sum_{k=1}^{n} P_u^k(y)$ is:

$$\hat{P}_u(y) = \frac{1}{2r_0} \sum_{k=1}^{n} \frac{1}{\sigma_k^2}[(\mu_k - A_k^Ty)_+]^2 \tag{5.2}$$

where $a_+ = \max(0,a)$.

Proof: We have to show that:

$$P_u^k(y) = \frac{1}{2r_o \sigma_k^2} [(\mu_k - A_k^T y)_+]^2 \tag{5.3}$$

Now from Proposition 5.1 we have for the k-th constraint:

$$P_u^k(y) = \sup_{x_k \geq 0} \{x_k (\mu_k - A_k^t y) - r_o \sigma_k^2 x_k^2\}$$

which by simple calculus gives (5.3).

□

REFERENCES

[1] Arrow, K.J.: Essays on the Theory of Risk-bearing, Markham Publishing, Chicago, 1975.

[2] Bamberg, G. and Spremann, K.: "Implications of constant risk aversion", Zeit. für Operations Research, 25, 1981, pp. 205-224.

[3] Cass, D. and Stiglitz, J.: "The structure of investor preferences and asset returns and separability in portfolio allocation: A contribution to the pure theory of mutual funds". J. Eco. Th., 2, 1970, pp. 122-160.

[4] Charnes, A. and Cooper, W.W.: "Chance constrained programming", Manag.Sci., 5, 1959, pp. 73-79.

[5] Hammond, J.S.: "Simplifying the choice between uncertain prospects where preference is non-linear", Manag.Sci., 20, 1974, pp. 1047-1072.

[6] Pratt, J.W.: "Risk aversion in the small and in the large", Econometrica, 32, 1964, pp. 122-136.

[7] Roberts, A.W., Varberg, D.E., Convex Functions, Academic Press, New York, 1973.

[8] von Neuman, J. and Morgenstern, O.: Theory of Games and Economic Behavior, 2nd ed., Princeton University Press, Princeton, N.J., 1947.

[9] Wilson, R.: "The theory of syndicates", Econometrica, 36, 1968, pp. 119-132.

A FEASIBLE SOLUTION TO DYNAMIC TEAM PROBLEMS
WITH A COMMON PAST AND APPLICATION
TO DECENTRALIZED DYNAMIC ROUTING

G. Casalino, F. Davoli, R. Minciardi, R. Zoppoli

Istituto di Elettrotecnica, Università di Genova
Viale F. Causa, 13 - 16145 Genova, Italy

ABSTRACT

Decentralized dynamic routing in communication networks is dealt with
in the paper. A procedure to find the optimal solution over a finite control horizon is given and then the structure of the optimal control law
is investigated. Finally, a receding horizon control scheme is proposed
whose application is made possible by the existence of a sufficient statistic.

1. INTRODUCTION

A classical example of decentralized (team) control problem deals with
routing in packet-switched computer communication networks. Several approaches have been attempted to determine satisfactory network routing protocols. Only recently, however, a solution to the routing problem has been
sought in a control-theoretic framework. In this case, the whole network
is modelled as a dynamic system, and closed-loop or "dynamic" routing strategies are determined.
The first important results in this direction are in the works of Segall
/1/ and Moss and Segall /2/, where a state-space model is introduced and
a closed-loop optimal policy is developed. Unfortunately, their approach
deals only with centralized feedback strategies, where the optimal control
(i.e., the routing actions) is, at each instant, a function of the overall
network state. This fact makes the approach proposed by Moss and Segall
practically infeasible, since it is very hard to assume that a single node
or a network routing centre may have, at each instant, the overall network
state information without any time delay. On the contrary, a more realistic
hypothesis is that the routing actions be determined at each instant only
on the basis of local information. A dynamic decentralized routing strategy
has been recently proposed by Sarachik and Özgüner /3/ whose approach is
that of forcing every node to apply a routing strategy which is optimal
for the routing problem in a network consisting of a single source node
and several parallel links leading to the unique destination. Another dy-

namic decentralized solution has been proposed by the authors /4/, which however yields only suboptimal strategies.

In this paper, the decentralized dynamic routing problem is considered in the framework of team control problems /5/. Much of the work previously done by the authors /6,7/ is used. The paper is organized as follows. In the next Section the basic problem is stated and a solution procedure is developed. In the third Section, the structure of the optimal control strategy is determined. Finally, in the fourth Section, a control scheme is introduced which resembles the so-called "receding-horizon" control. This scheme represents a possible feasible approach for decision processes extending over an infinite number of stages and for which no optimal stationary solution over infinite horizon is known.

2. STATEMENT OF THE CONTROL PROBLEM AND ITS SOLUTION

Consider a network with N nodes and L directed links. $M \leqslant N$ nodes are traffic destinations and will be denoted by $d_1, \ldots, d_M$. The time variable is discrete. We make the following assumption, for sake of simplicity: the traffic overhead due to the conveyance of information about the network state (queue lengths) is in all cases negligible with respect to the traffic which conveys information which has to be transmitted by the network. This assumption has not been made in /1/. However, it by no doubt makes the problem tractable and is not unrealistic in many practical cases. We suppose that the network topology is time-invariant, so that the only stochastic events in the network are the traffic inputs. Then, let $r_i^{(d)}(t)$ be the input traffic (measured in some convenient unit) that enters the network at node i, with node d as destination, in the time slot (t,t+1), $u_{ij}^{(d)}(t)$ be the traffic destinated to node d, passing through the directed link (i,j) during the same interval, and finally $x_i^{(d)}(t)$ the amount of traffic destinated to node d, which at time t is queued at node i. Thus, the state equations of the overall system are

$$x_i^{(d)}(t+1) = x_i^{(d)}(t) - \sum_{l \in S_i} u_{il}^{(d)}(t) + \sum_{k \in P_i} u_{ki}^{(d)}(t) + r_i^{(d)}(t) \qquad (1)$$

for t=0,1,..., i=1,...,N, $d=d_1,\ldots,d_M$, and with $d \neq i$. P_i is the set of nodes predecessors of node i, whereas S_i is the set of nodes successors of node i. For sake of simplicity, no processing times are considered either in the input or in the intermediate nodes. The external traffic inputs $r_i^{(d)}(t)$, i=1,...,N, $d=d_1,\ldots,d_M$, are modelled as discrete-time discrete-valued (integer) stochastic processes, and the initial states $x_i^{(d)}(o)$, i=1,...,N, $d=d_1,\ldots,d_M$ are discrete-valued (integer) random variables. Variables $x_i^{(d)}(t)$ and $u_{ij}^{(d)}(t)$ are also integer. Taking into account also capacity constraints, we have altogether

$$x_i^{(d)}(t) \geqslant 0 \qquad\qquad \forall i,\ t,\ d \neq i \qquad\qquad (2a)$$

$$u_{ij}^{(d)}(t) \geqslant 0 \qquad\qquad \forall i,\ t,\ d \neq i,\ j \in S_i \qquad\qquad (2b)$$

$$\sum_{d \neq i} u_{ij}^{(d)}(t) \leqslant C_{ij} \qquad\qquad \forall i,\ t,\ j \in S_i \qquad\qquad (2c)$$

being C_{ij} the capacity of link (i,j). These constraints must be satisfied for all realizations of the primitive random variables (inputs and initial states).

The control objective is that of minimizing the functional

$$J = E \left[\sum_{t=0}^{T-1} \sum_{i=1}^{N} \sum_{d \neq i} w_{id}\, x_i^{(d)}(t+1) \right] \qquad\qquad (3)$$

where the expectation is taken over the probability space of the inputs and initial states random variables, and w_{id} are weighting coefficients. If all w_{id}'s are equal to 1, then the functional (3) is simply the aggregate delay in the network up to time t. Let $x_i(t) \triangleq col(x_i^{(d)}(t),\ d \neq i)$, $u_i(t) \triangleq col(u_i^{(d)}(t),\ d \neq i)$, being $u_i^{(d)}(t) \triangleq col(u_{ij}^{(d)}(t),\ j \in S_i)$. The admissible control strategies are informationally decentralized, and are of the form $u_i(t) = f_{i,t}(I_i(t))$, $i=1,\ldots,N$, $t=0,1,\ldots,T-1$, where $I_i(t)$ is the information set of decision maker $DM_i(t)$ located at node i at time instant i.

Let us now define the information structure as follows. $DM_i(t)$ knows "its own state component" $x_i(t)$, and, moreover, it acquires the information about the state components corresponding to all other nodes in the network with a number of time instants of delay equal to the number of links (anyhow oriented) corresponding to the topological distance from node i. This definition clearly requires each node to transmit the "new" components of its network state information to all its neighbours within each control interval. Of course, we suppose that each node has a perfect memory of all the network state information it has ever acquired. Summing up, we have

$$I_i(t) \triangleq \left\{ x_i(t'),\ t' \leqslant t \right\} \bigcup \left\{ x_j(t'-k_{ij}),\ \text{for } j \neq i,\ \text{with } k_{ij} = \text{topological di-}\right.$$

stance between nodes i and j, provided $k_{ij} \leqslant t'$, $t' \leqslant t \Big\}$ (4)

Of course, we can define also an information vector $z_{i,t}$, whose components correspond to the elements of $I_i(t)$. Note that $I_i(t)$ does not include stochastic inputs $r_i^{(d)}(t)$, but that all stochastic inputs $r_i^{(d)}(\tau)$ with $\tau < t$, can be deduced from $I_i(t)$ and from the knowledge of the control laws. This means that the external inputs cannot be directly measured from the decision maker.

The above introduced information structure is linear in the primitive random variables (inputs and initial states), and, besides, it turns out to be partially nested /5/, which, shortly speaking, means that if a decision agent influences another one, then the information set of the former is necessarily included in the information set of the latter agent. The above properties imply the possibility, for each agent of the team, of re-constructing the control actions of the agents which affect its own informa-tion set.

It is apparent that, by use of state equations (1), the cost (3) to be minimized can be expressed also in the form

$$J = E \left[\sum_{t=0}^{T-1} \sum_{i=1}^{N} a_{i,t} \ u_i(t) + \text{terms depending only on the primitive r.v.'s} \right] \tag{5}$$

where $a_{i,t}$ are properly defined row vectors. Also constraints (2a) can be expressed in a form involving only primitive random variables and control actions, namely

$$x_i^{(d)}(o) + \alpha_{i,t}^{(d)} u_o^{t-1} + \beta_{i,t}^{(d)} r_o^{t-1} \geq 0 \qquad \forall i,t, \qquad d \neq i \tag{6}$$

where $\alpha_{i,t}^{(d)}$ and $\beta_{i,t}^{(d)}$ are row vectors determined by the network topolo-gy, $u_o^{t-1} \triangleq \text{col}(u(o),\ldots,u(t-1))$, being $u(t) \triangleq \text{col}(u_1(t),\ldots,u_N(t))$, and r_o^{t-1} is similarly defined.

At this point, it is worth recalling that, due to the above hypotheses about the information structure of the team, it is possible /5/ to find an information structure which is equivalent to the above introduced one, and where the information vector $\hat{z}_{i,t}$ of $DM_i(t)$ is a (linear) function only of the primitive random variables, that is

$$\hat{z}_{i,t} = H_{i,t} \xi \tag{7}$$

where $\xi \triangleq \text{col}(x(o), r_o^{T-1})$ and the definition of vector $x(o)$ is straight-forward. Assume now that all the primitive random variables are constrained within a certain range. Since these variables are integer, then the number of possible values of ξ is finite. Then, also the set of possible values of each information vector $\hat{z}_{i,t}$ is finite, say, $\hat{z}_{i,t}^1,\ldots,\hat{z}_{i,t}^{\lambda_{i,t}}$. Let $u_i^s(t)$ be the value of the control (vector) action of $DM_i(t)$ which corresponds to the value $\hat{z}_{i,t}^s$ of $\hat{z}_{i,t}$. The vector $\bar{u} = \text{col}[u_i^s(t), \ s=1,\ldots, \lambda_{i,t}, \ \forall i; \ \forall t]$ col-lects all possible values of the control vectors for each node at each in-stant. Let moreover $p_{i,t}^s$ be the probability of occurrence of the value $\hat{z}_{i,t}^s$ (that, of course, can be easily computed from the p.d.f. of the primitive random variables). Then the minimization of (5) is equivalent to the minimi-zation of

$$J_2 = \sum_{t=0}^{T-1} \sum_{i=1}^{N} \sum_{s=1}^{\lambda_{i,t}} a_{i,t} \ u_i^s(t) \ p_{i,t}^s \tag{8}$$

where the inner summation is carried over all possible values of control vector $u_i(t)$. Note that the decentralization of the control strategies is taken into account. More specifically, if, for different realizations ξ_a

and ξ_b of the random vector ξ, the static information vector $\hat{z}_{i,t}$ assumes the same value, i.e., $H_{i,t} \xi_a = H_{i,t} \xi_b$, then the control vectors corresponding to such realization, namely $u_o^{T-1}(\xi_a)$ and $u_o^{T-1}(\xi_b)$, must be identical in their components corresponding to vector $u_i(t)$. Constraints (2b), (2c) and (6) can easily be converted into a form suitable for the minimization of (8) (i.e., converted into constraints over values $u_i^s(t)$). More specifically, constraints (2b) and (2c) become respectively

$$u_{ij}^{(d),s}(t) \geqslant 0 \qquad\qquad s=1,\ldots,\lambda_{i,t}, \quad \forall i,t,d\neq i,j\in S_i \qquad (9)$$

$$\sum_{d\neq i} u_{ij}^{(d),s}(t) \leq c_{ij} \qquad\qquad s=1,\ldots,\lambda_{i,t}, \quad \forall i,t,j\in S_i \qquad (10)$$

where obviously $u_{ij}^{(d),s}(t)$ is a component of vector $u_{ij}^s(t)$.

The convertion of (6) requires a little more attention. In fact, consider the finite number of possible values of the stochastic vector $\xi(t)$ which is defined as $\xi(t) \triangleq col(x(o), r_o^{t-1})$ if $t > 0$, and $\xi(o) \triangleq x(o)$. Let these possible values be $\xi_1(t),\ldots, \xi_{Q(t)}(t)$. Now let Λ_{t-1}^m a matrix defined so that $\Lambda_{t-1}^m \bar{u}$ is the vector collecting the values of the control actions that are taken by the various agents of the team from instant o up to instant (t-1), when realization $\xi_m(t)$ of $\xi(t)$ occurs. Of course, the structure of the matrix Λ_{t-1}^m is strictly related to the (static) information structure of the team. Then, it is readily seen that constraints (6) can be replaced by constraints

$$[e \mid \alpha_{i,t}^{(d)}] \xi_m(t) + \alpha_{i,t}^{(d)} \Lambda_{t-1}^m \bar{u} \geqslant o \qquad\qquad (11)$$

$$m=1,\ldots,Q(t), \quad \forall i,t,d\neq i$$

where e is a row vector of all zeroes but with a single one in a suitable position. It is worth observing that many of the constraints (11) may turn out to be redundant and can be eliminated. To be clearer, consider, for instance, for an arbitrary network, the following constraint

$$x_2^{(d)}(o) + u_{12}^{(d)}(o) - u_{23}^{(d)}(o) - u_{24}^{(d)}(o) + r_2^{(d)}(o) \geqslant o \qquad\qquad (12)$$

which corresponds to $x_2^{(d)}(1) \geqslant 0$. Then, since $r_1^{(d)}(o)$ and $r_2^{(d)}(o)$ cannot be deduced from the information set $I_1(o)$ nor from $I_2(o)$, two realizations of $\xi(1)$ which are equal in the value of $x_2^{(d)}(o)$ must give rise to the same values of the control actions $u_{12}^{(d)}(o)$, $u_{23}^{(d)}(o)$, $u_{24}^{(d)}(o)$. Writing the constraints (11) corresponding to constraint (12) gives many identical constraints (two of these constraints are identical if they correspond to two realizations of $\xi(1)$ which do not differ in the value of $x^{(d)}(o)$ and $r_2^{(d)}(o)$). Even eliminating this redundancy, still many constraints are not necessary, since, among all constraints corresponding to realizations of $\xi(1)$ which differ only in the value of $r_2^{(d)}(o)$, it is sufficient to keep only the most restrictive, i.e., the one corresponding to the value $r_2^{(d)}(o) = 0$. Notwithstanding these observations, we shall continue for sake of simplicity, to use the notation (11) to express the constraints corresponding to (6), even if care must be used in every particular problem to obtain the actual set of effective constraints.

At this point, we can conclude that the determination of the optimal control actions $u_{ij}^{(d),s}(t)$ can be performed by solving an integer linear programming problem whose objective function is given by (8), with constraints (9), (10) and (11), besides the integrity constraint over the decision variables. Of course, the optimal control strategies (functions of the static information vectors $\hat{z}_i(t)$) are determined in a tabular form. Thus, decision maker $DM_i(t)$ has first to convert his dynamic information vector $z_i(t)$ into $\hat{z}_i(t)$ and then to apply the optimal control strategy. In the above conversion, the knowledge of the other agents' strategies (and then of all the decision tables in the network) is required, if control actions are not exchanged between nodes. Thus, to avoid heavy storage requirements at each node, we suppose that the nodes exchange not only the state information but also the control action information between each other, and that each node has a perfect memory of its past control actions. This assumption does not yield great communication requirements and modifies in a straightforward way the above given definition (4) of the information sets $I_i(t)$. A final observation is needed as to the reduction of the dynamic team problem into a static one. Actually, this reduction is not strictly necessary, since the probabilities $p_{i,t}^s$ in (8) could be computed also with a dynamic information structure (due to the partially nestedness hypothesis). In the same way, also constraints of the type of (11) could be easily written even with a dynamic information structure. Then, summing up, the above considered reduction is performed only for sake of simplicity, but the only property actually necessary is the partially nestedness.

3. STRUCTURE OF THE OPTIMAL CONTROL STRATEGY

The solution procedure proposed in the previous Section suffers from the following serious drawback. The dimensionality of the integer linear programming problem whose solution is needed to obtain the optimal control laws, readily increases, as is easy to see, with the control horizon. This makes almost impossible to use this solution procedure in case of "long" control horizons. However, it is clear that we are actually interested in long (possibly infinite) horizon control problems. To circumvent such difficulty, a receding-horizon control scheme can be applied, as will be detailed in the next Section. To this end, a preliminary investigation is necessary   about the structure of the optimal control strategy.

Let us observe that, besides   the linearity and partially nestedness properties, the information structure considered in the previous Section is characterized by a third important property, namely the existence of a common past information set. More specifically, if we consider the maximum topological distance between nodes in the network, say k, for time instants t ≥ k, the information set

$$
I(t-k) \triangleq \left\{ x_i(\tau), u_i(\tau), i=1,\ldots,N, \quad 0 \leq \tau \leq t-k \right\} \tag{13}
$$

is known to all agents $DM_i(t')$, with $t' \geq t$. Note that the exchange of control actions is assumed, as previously indicated.

Consider, at this point, the following auxiliary problem consisting in the minimization of the cost

$$
J^t = E\left[\sum_{\tau=t}^{T-1} \sum_{i=1}^{N} \sum_{d \neq i} w_{id} x_i^{(d)}(\tau+1) \,|\, I(t-k) \right] \tag{14}
$$

considering the strategies at time instants $0,\ldots,t-1$ fixed and known for all nodes. Here again, by use of state equations (1), an alternative expression can be found for J_1^t, namely

$$
J_1^t = E\left[\sum_{\tau=t}^{T-1} \sum_{i=1}^{N} a_{i,\tau}^t u_i(\tau) + \text{terms depending only on the primitive} \right. \\
\left. \text{r.v.'s and on vector } u_{t-k}^{t-1} \right] \tag{15}
$$

The cost (15) must be minimized with constraints (2b), (2c) (with t replaced by $\tau \geq t$) and, in lieu of (6),

$$
x_i^{(d)}(t-k) + \alpha_{i,\tau}^{-t,(d)} u_{t-k}^{t-1} + \overline{\overline{\alpha}}_{i,\tau}^{t,(d)} u_t^{\tau-1} + \beta_{i,\tau}^{t,(d)} r_{t-k}^{\tau-1} \geq 0 \tag{16}
$$

$$
\forall i, \ \tau > t, d \neq i
$$

where all symbols have an obvious meaning. The control actions collected by the vector u_{t-k}^{t-1} are supposed to be fixed and known functions of the personal information set of the corresponding decision makers. By expressing all variables appearing in the personal information sets and corresponding to time interval $[(t-k),(t-1)]$, as functions of $x(t-k)$ and r_{t-k}^{t-2}, we can obtain another expression of u_{t-k}^{t-1}, namely

$$
u_{t-k}^{t-1} = \tilde{f}_{t-k}^{t-1}(I(t-k), r_{t-k}^{t-2}) \tag{17}
$$

We can define, for any fixed realization of $I(t-k)$ the corresponding "restriction" of the vectorial function $\tilde{f}^{t-1}_{t-k}$, namely

$$u^{t-1}_{t-k} = \tilde{f}^{t-1}_{t-k} \, (I(t-k), \, r^{t-2}_{t-k}) \triangleq \gamma^{t-1}_{t-k, \, I(t-k)} \, (r^{t-2}_{t-k}) \tag{18}$$

The vectorial function $\gamma^{t-1}_{t-k, \, I(t-k)}$ is expressed by a set of tables. Let us define a vector δ^{t-1}_{t-k} which collects the ordered values of the elements of these tables. Clearly, δ^{t-1}_{t-k} is a function of $I(t-k)$.

In order to solve the auxiliary problem, a procedure quite similar to the one presented in the previous Section can be followed. First of all, the dynamic information structure of the team is converted to a static one. Namely, consider the information set of $DM_i(\tau)$, with $\tau \geqslant t$,

$$I_i(\tau) = \left\{ I(t-k), \, z^t_{i,\tau} \right\} \tag{19}$$

where the definition of $z^t_{i,\tau}$ is straightforward. Of course, $z^t_{i,\tau}$ is a linear function of $x(t-k)$, of $r^{\tau-1}_{t-k}$ and of $u^{\tau-1}_{t-k}$. Due to the partially nestedness and to the linearity property, the above information structure is equivalent to the following one

$$\hat{I}_i(\tau) = \left\{ I(t-k), \, \hat{z}^t_{i,\tau} \right\} \tag{20}$$

where

$$\hat{z}^t_{i,\tau} = H^t_{i,\tau} \, \eta^t \tag{21}$$

having defined $\eta^t \triangleq r^{T-1}_{t-k}$ for convenience of notation. The solution can now follow the same lines as in the previous Section. Remind that the whole auxiliary problem is conditioned to a certain fixed realization of $I(t-k)$. Thus, vectors $x(t-k)$ and δ^{t-1}_{t-k} are to be considered fixed and known. Note that the knowledge of δ^{t-1}_{t-k} implies the knowledge of the function of r^{t-2}_{t-k} defined in (18). Then, the control strategies which solve the auxiliary problem, for the particular $I(t-k)$ fixed, can be found by minimizing the cost

$$J^t_1 = \sum_{\tau=t}^{T-1} \sum_{i=1}^{N} \sum_{s=1}^{\lambda^t_{i,}} a^t_{i,\tau} \, u^{t,s}_i(\tau) \, p^{t,s}_{i,\tau} \tag{22}$$

with the constraints

$$u_{ij}^{(d),s}(\tau) \geqslant 0 \qquad\qquad s=1,\ldots,\lambda_{i,\tau}^t, \qquad \forall\, i,\tau\,, d\neq i, j \in S_i \qquad\qquad (23)$$

$$\sum_{d\neq i} u_{ij}^{(d),s}(t) \leq C_{ij} \qquad\qquad s=1,\ldots,\lambda_{i,\tau}^t, \qquad \forall\, i,\tau\,, j\in S_i \qquad\qquad (24)$$

and

$$x^{(d)}(t-k) + \beta_{i,\tau}^{t,(d)}\eta_m^t(\tau) + \overline{\alpha}_{i,\tau}^{t,(d)}\varphi_{t-k,I(t-k)}^{t-1}(\eta_m^t(\tau))$$

$$+ \overline{\overline{\alpha}}_{i,\tau}^{t,(d)} u_t^{\tau-1}(I(t-k),\eta_m^t(\tau)) \geqslant 0 \qquad\qquad (25)$$

$$m=1,\ldots,Q_t(\tau),\forall\, i,\tau\,, d\neq i$$

where the symbols have an analogous meaning to that of symbols in (8)–(11). Namely, $\eta^t(\tau) = r_{t-k}^{\tau-1}$ has $Q_t(\tau)$ possible realizations; $\varphi_{t-k,I(t-k)}^{t-1}(\eta_m^t(\tau))$ are the control actions corresponding to the function (18) (i.e. to vector δ_{t-k}^{t-1})and to the realization $\eta_m^t(\tau)$ of $\eta^t(\tau)$; $u_t^{\tau-1}(I(t-k),\eta_m^t(\tau))$ is a collection of control values $u_i^{t,s}(\tau)$ (formed according to the static information structure of the team) and are to be determined. $\lambda_{i,\tau}^t$ is the number of possible different realizations of $z_{i,\tau}^t$, in correspondence of which the various control actions $u_i^{t,s}(\tau)$ are applied. The probabilities $p_{i,\tau}^{t,s}$ can be again easily computed on the basis of a priori statistical information.

It is apparent from above that the knowledge of I(t-k) affects the integer linear programming problem defined by cost (22) and constraints (23)--(25), only through vectors x(t-k) and δ_{t-k}^{t-1}. More specifically, the knowledge of these vectors is needed in building the constraints above summarized by (25). Thus, these vectors together represent the "contraction" of the common past information set which is necessary to find the control strategies between t and (T-1). By solving a linear integer programming problem of the above type for all possible values of vector $[x(t-k), \delta_{t-k}^{t-1}]$, one solves the auxiliary problem completely (note that the values of this vector are finite in number, due to capacity constraints and to input range limitations). The above discussion can be summarized in the following

Lemma. The optimal strategies which solve the auxiliary problem are of the form

$$u_i(\tau) = g_{i,\tau}^t (x(t-k),\, \delta_{t-k}^{t-1},\, r_{t-k}^{\tau-1}) \qquad\qquad \forall\, i,\tau =t,\ldots,T-1 \qquad (26)$$

$\square$

We are now in a position to state the following basic result.

Theorem. Consider the optimization problem corresponding to the minimization of (3) with constraints (2). Then, for every instant t such that $k \le t \le T-1$, the optimal control strategy has the structure

$$u_i(t) = g_{i,t}^t \ (x(t-k), \delta_{t-k}^{t-1}, r_{t-k}^{t-1}) \qquad \forall i \qquad (27)$$

The vectors $x(t-k)$ and δ_{t-k}^{t-1} condense all past information, up to time instant $(t-k)$ which is necessary to determine the control actions. They have constant dimensions and can be recursively updated on the basis of their previous values and of $u(t-k)$. Thus, the collection of $x(t-k)$, δ_{t-k}^{t-1} and $u(t-k)$ plays the role of a sufficient statistic. $\qquad\qquad$ □

Proof. In order to prove the theorem, it is necessary first to fix the control strategies between time instant o and $(k-1)$ to be the optimal ones (which, of course, are determined solving the whole optimization problem in the way indicated in the previous Section). Then, an auxiliary problem (a.p.) can be posed, for t=k, and strategies (26), with t=k, i.e.

$$u_i(\tau) = g_{i,\tau}^k \ (x(o), \delta_o^{k-1}, r_o^{\tau-1}) \qquad \forall i, \ \tau = k,\ldots,T-1 \qquad (28)$$

are actually optimal for the whole problem. At this point, a new auxiliary problem can be considered, consisting in the minimization of

$$J^{k+1} \triangleq E \ [\sum_{\tau=k+1}^{T-1} \ \sum_{i=1}^{N} \ \sum_{d \ne i} \ w_{id} \ x_i^{(d)}(\tau+1)|I(1)] \qquad (29)$$

with constraints (2b), (2c) (with t replaced by $\tau \ge (k+1)$) and (16), written for t=k+1, and given the strategies previously fixed (as the optimal ones) up to time instant $(k-1)$, and the strategy at time instant k just obtained by solving the previous a.p.. Clearly, the strategies which solve the new a.p. are a part of those determined by solving the previous a.p. (that is, the strategies given by (28), for τ=k+1,...,T-1). However, another form of these strategies can be found by solving the new a.p.. Namely, due to the perfect similarity of the two a.p.'s, they can also be expressed as

$$u_i(\tau) = g_{i,\tau}^{k+1} \ (x(1), \delta_1^{k-1}, r_1^{\tau-1}) \qquad \forall i, \ \tau = k+1,\ldots,T-1 \qquad (30)$$

Then, by considering successive a.p.'s and picking up only the first of the strategies which solve these problems, the expressibility of the optimal strategies as given by (27) is readily proved. $\qquad\qquad$ To complete the proof of the theorem, it remains to show the recursive computability of vectors $x(t-k)$ and δ_{t-k}^{t-1}. Clearly, no problem exists for vector $x(t-k)$, which is actually included in the common information set $I(t-k)$, whereas some care is necessary to show the updating law of δ_{t-k}^{t-1}.

To this end, let us begin from time instant $t=k$. In this instant $I(t-k) = I(o) = \{x(o), u(o)\}$. By solving the auxiliary problem between $t=k$ and $(T-1)$, the strategies (28) are obtained. The control strategies up to time instant $(k-1)$ have been fixed to be the optimal ones and are

$$u_o^{k-1} = \tilde{f}_o^{k-1}(I(o), r_o^{k-2}) \tag{31}$$

Then, for any possible realization of $I(o)$, the function

$$u_o^{k-1} = \gamma_{o,I(o)}^{k-1}(r_o^{k-2}) \tag{32}$$

is fixed. As above indicated, each of these functions can be represented by a table, whose entries are collected in a finite-dimensional vector, namely δ_o^{k-1}.

Let us consider now time instant $t=k+1$. In this instant, the optimal strategy can be expressed as

$$u_i(k+1) = g_{i,k+1}^{k+1}(x(1), \delta_1^k, r_1^k) \tag{33}$$

What we want to show is that δ_1^k can be computed on-line on the basis of δ_o^{k-1} and on-line information. To this end, note that vector δ_{t-k}^{t-1} is given by

$$\delta_{t-k}^{t-1} = \text{col}(\delta_{t-k}(t-k), \ldots, \delta_{t-k}(t-1)) \tag{34}$$

where $\delta_{t-k}(j)$ collects all entries of the table corresponding to the function $\gamma_{j,I(t-k)}(r_{t-k}^{j-1})$, which is the j-th function of the collection $\gamma_{t-k,I(t-k)}^{t-1}(r_{t-k}^{t-2})$ defined by (18).

At this point, observe that the function $\gamma_{j,I(1)}(r_1^{j-1})$, for $1 \le j \le k-1$ can be determined (on-line) as the restriction of the function $\gamma_{j,I(o)}(r_o^{j-1})$ performed by fixing the value of $r(o)$ which can be deduced from $x(o)$, $u(o)$ and $x(1)$, via the state equations. What above stated is equivalent to say that it is possible to obtain $\delta_1(j)$ from $\delta_o(j)$ and on-line information, for $1 \le j \le k-1$. Thus, it remains to compute $\delta_1(k)$. Then, consider the first strategy of (28), namely

$$u_i(k) = g_{i,k}^k(x(o), \delta_o^{k-1}, r_o^{k-1}) \qquad \forall i \tag{35}$$

We are at time instant $(k+1)$. Thus $x(o)$, δ_o^{k-1} are known to all nodes and r_o is computable. By fixing these variables to their values in (35), one clearly obtains the function $\Psi_{k,I(1)}(r_1^{k-1})$, or equivalently the vector $\delta_1(k)$ collecting all entries of the corresponding table.

Summing up, we have proved that

i) $\delta_1(j)$ is computable on the basis of $\delta_o(j)$, $x(o)$, $u(o)$, $x(1)$, $1 \leq j \leq k-1$

ii) $\delta_1(k)$ is computable on the basis of $x(o)$, δ_o^{k-1}, $u(o)$, $x(1)$.

Thus, the whole vector δ_1^k is computable on the basis of δ_o^{k-1}, $u(o)$, $x(1)$. Clearly, the same arguments can also be applied to show that δ_{t-k+1}^t is computable on the basis of δ_{t-k}^{t-1}, $u(t-k)$, $x(t-k+1)$. Moreover, the above discussion shows also that the initialization of this computation, i.e. the determination of δ_o^{k-1} is simply performed by picking, at time instant k, the function (32) (and then the vector δ_o^{k-1}) corresponding to the actually realized $I(o)$. Thus, the proof of the theorem is complete. $\triangle$

Remark. The main result given by the previous theorem is the determination of the structure of the optimal strategies and the proof of the existence of a sufficient statistic. The theorem states that for $t \geq k$ the form of the optimal strategy remains unaltered and the tables which "store" these strategies have constant dimensions. The whole determination of strategies (27) can be performed off-line. This requires: i) the solution of the problem from o to $(T-1)$ to determine the control strategies corresponding to the first k instants; ii) the solution of $(T-k)$ auxiliary problems (from t=k to t=T-1). Note that, for the solution of each auxiliary problem, the integer linear programming problem corresponding to the minimization of (22), with constraints (23)-(25), must be solved for any possible value of vector $[x(t-k), \delta_{t-k}^{t-1}]$. Due to this fact, the off-line computational requirements may turn out to be rather high, and an alternative possibility of applying the above result may become convenient. This second possibility corresponds to determining off-line the first k optimal strategies (by solving the whole optimization problem) and then to determine on-line the optimal strategies (27) for successive time instants. The advantage is given in this case by the reduction of the computations necessary to find strategies (27). Namely, to find, for each time instant, this expression of the optimal control strategy, it is necessary to solve only the integer linear programming problem corresponding to the minimization of (22), with constraints (23)-(25), only for the particular realization of vector $[x(t-k), \delta_{t-k}^{t-1}]$ which has actually been recognized on-line. Of course, this operating procedure makes sense only if the on-line computational requirements do not become prohibitive. $\square$

4. A RECEDING HORIZON CONTROL SCHEME

In this Section we shall consider a possible interesting use of the results obtained in the previous Section. Suppose to have solved the auxiliary

problem corresponding to time instant t, and then consider the following
problem

$$
\min \quad \tilde{J}^{t+1} = E\left[\sum_{\tau=t+1}^{T} \sum_{i=1}^{N} \sum_{d \neq i} w_{id} \, x_i^{(d)}(\tau+1) \,\middle|\, I(t-k+1)\right]
\tag{36}
$$

with constraints (2b), (2c) (with t replaced by $\tau \geqslant$ t+1) and (16) (with t
replaced by (t+1)), given the strategies corresponding to the last k time
instants. Here again, due to the similarity of the two problems, we can
state that the optimal control strategy for time instant (t+1) has the struc‑
ture

$$
u_i(t+1) = g_{i,t+1}^{t+1} (x(t-k+1), \, \delta_{t-k+1}^{t}, \, r_{t-k+1}^{t}) \qquad \forall i
\tag{37}
$$

where vectors x(t-k+1) and δ_{t-k+1}^{t} can be computed on-line by the same sim‑
ple updating mechanism as in the previous Section. Going further, we can
pose and solve another optimization problem consisting in minimizing a cost
functional identical to (36) but for (t+1) replaced by (t+2) and T replaced
by (T+1), and so on. Clearly, the only difference with respect to the sequen‑
ce of auxiliary problems considered in the previous Section is that, in the
present case, any of the optimization problems which are subsequently solv‑
ed is posed over an interval of constant length .
    The proposed control scheme exhibits close connections with the concept
of the so-called "receding-horizon" control laws. This term derives from
the fact that the controller sees an apparent terminal time which is always
at the same distance in the future. The receding-horizon notion has proved to
be an efficient tool for designing stable state-feedback controllers for
time-varying linear and nonlinear systems.   Observe that the receding-
-horizon criterion naturally moves our original problem into the class of
the decision processes characterized by an infinite number of stages. Clear‑
ly, for the applicative problem we are dealing with, this is a positive
fact, since it is more realistic to assume that the process does not end
at a given time instant, but goes on for an indefinite number of decision
steps. Moreover, the time window of each optimization problem can be chosen
appropriately, so as to obtain a reasonable compromise between the computa‑
tional complexity of the problem (increasing with the control horizon
length) and the physical properties of the system.
    To apply the receding-horizon control scheme, it is necessary, in the
present case, to initialize   the procedure by arbitrarily fixing the con‑
trol strategies for time instants from 0 up to (k-1). Then, for every time
instant, an optimization problem similar to the auxiliary problem described
in the previous Section must be solved. Clearly, the control strategies are,
after the first k instants, time-varying, but with a constant structure.
An off-line computation of these strategies can hopefully show that these

time-varying strategies become, under suitable hypotheses, eventually statio nary. This point is actually matter of investigation. Alternatively, an on--line implementation of the above receding-horizon scheme can be conceived, where the time-varying strategies, for time instants $t \geqslant k$, are determined by solving on-line optimization problems which are much simpler that the whole auxiliary problems (in that the value of vector $[x(t-k), \delta_{t-k}^{t-1}]$ is known).

5. CONCLUDING REMARKS

In this paper we have proposed a possible approach to face the dynamic decentralized routing in computer communication networks. This approach is made possible by several hypotheses, most of which are related to the information structure of the team control problem.

The main result of the paper is the proof of existence of a sufficient statistic, which allows the definition of a receding-horizon control scheme.

Generally speaking, the application of a receding-horizon scheme in a team control environment requires the solution of a sequence of team problems characterized by a fixed number of decision agents. However, each problem is a dynamic one and its solution may involve formidable difficulties. On the whole, the following assumptions are needed to allow the application of a receding-horizon control procedure.

i)   The team information structure must be partially nested. For linear information structures, this assumption enables a dynamic team optimization problem to be reduced to a static one.

ii)   The static team problem must be "appropriately structured" in order that it can be resolvable. Two cases exhibit such an appropriate structure: 1) LQG optimization problems /8/, 2) linear programming problems under uncertainty, provided that the primitive random variables are discrete and take on a finite number of values (the problem considered in this paper falls into this class of team optimization problems).

iii) The information structure of the team must allow the definition of a common past information set $I(t-k)$.

iv)   It must be possible to contract the information set $I(t-k)$ into a sufficient statistic characterized by: 1) a time-invariant dimension, 2) the possibility of being determined recursively through a procedure involving a time-invariant computational effort.

REFERENCES

/1/ A. Segall, The modeling of Adaptive Routing in Data Communication Net-
    works, IEEE Trans. Comm., Vol. COM-25, pp. 85-95, 1977.

/2/ F.H. Moss and A. Segall, An Optimal Control Approach to Dynamic Routing
    in Networks, IEEE Trans. Autom. Control, Vol. AC-27, pp. 329-339, 1982.

/3/ P.E. Sarachik and U. Ozgüner, On Decentralized Dynamic Routing for Conge
    sted Traffic Networks, IEEE Trans. Autom. Control, Vol. AC-27, pp. 1233-
    -1238, 1982.

/4/ G. Bartolini, G. Casalino, F. Davoli, R. Minciardi and R. Zoppoli, A
    Team Theoretical Approach to Routing and Multiple Access in Data Communi
    cation Networks, Proc. of the 3rd IFAC/IFORS Symposium on Large Scale
    System, Warsaw, 1983.

/5/ Y.C. Ho and K.C. Chu, Team Decision Theory and Information Structures in Op-
    timal Control Problems - Part. I, IEEE Trans. Autom. Control, Vol. AC-
    -17, pp. 15-22, 1972.

/6/ G. Casalino, F. Davoli, R. Minciardi and R. Zoppoli, Decentralized Dyna-
    mic Routing in Data Communication Networks, Proc. Mediterranean Elet-
    trotechnical Conference, Athens, 1983.

/7/ G. Casalino, F. Davoli, R. Minciardi and R. Zoppoli, On the Structure
    of Decentralized Dynamic Routing Strategies, Proc. 22nd IEEE Conf. on
    Dec. & Control, San Antonio, Texas, pp. 472-476, 1983.

/8/ G. Casalino, F. Davoli, R. Minciardi and R. Zoppoli, Sufficient Stati-
    stics in Team Control Problems with a Common Past, Proc. 21st IEEE Conf.
    on Dec. & Control, Orlando, Florida, pp. 186-190, 1982.

STOCHASTIC CONSTRUCTION OF (q,M) PROBLEMS

M. Cirinà

Università di Torino

Dipartimento Scienze dell'Informazione

Corso d'Azeglio 42, Torino, Italy

This paper is concerned with the pseudorandom generation of linear complementarity problems that possess a solution and are not easy to solve.

1. Introduction

Consider the linear complementarity problem

(q,M) to find $x \in R^n$ satisfying $x \geq 0$, $Mx+q \geq 0$, $x^T(Mx+q)=0$,

where $q \in R^n$ and $M \in R^{n \times n}$ are given, vectors of R^n are written in column form and T denotes transposition. It is well known that this is an important problem in optimization theory and practice. For instance it includes as a special case the Karush-Kuhn-Tucker conditions for the standard quadratic program. On the other hand optimality conditions for equality and inequality constrained minimization problems constitute a subject initiated by the famous investigations of Lagrange and Fourier; see for instance Kuhn-Tucker [7], Prekopa [11].
The first important methods for solving (q,M) are contained in the papers of Cottle and Dantzig [5], and Lemke [8]. Subsequent research in linear complementarity remained oriented in part towards the enlargement

---

Work supported in part by Fondi Ministeriali per la Ricerca Scientifica (60%).

of the class of solvable (q,M) problems; for results in such direction see Chandrasekaran [1], Eaves [6], Mangasarian [9], Cirinà [2]. See also Van der Heyden [12] and Cottle [4].

In view of the fact that Lemke's complementary pivoting algorithm is a rather efficient one, e.g. Murty [10], it is reasonable that further research in the direction just mentioned should be concentrated mainly on (q,M) problems left unsolved by such algorithm.

This paper is concerned with the problem of generating randomly (q,M)'s that possess a complementary solution and are difficult to solve in the sense that they cannot be solved by Lemke's algorithm.

If we fix pseudorandomly a square matrix M of real numbers, there is no reason to believe that there exists a vector q such that (q,M) has a solution and it is difficult to solve. However under mild hypotheses on M (namely whenever M has at least two nonzero entries, one on the main diagonal and one off it) we prove (theorem 1) that by possibly changing the sign of some entries of M and interchanging some entries of two appropriate rows, M can be transformed into a matrix $\tilde{M}$ for which a vector q can be found so that $(q,\tilde{M})$ has the required property.

The proof of theorem 1 is constructive and can be used to write a routine for producing randomly linear complementarity problems with the property above. For one such routine see Cirinà [3] where it is made use of a (less simple) variant of the construction in theorem 1 with the purpose of keeping changes in M to a minimum.

The class of (q,M) problems thus generated is therefore operationally usable and it is useful for the orientation of new research as well as for testing new algorithms.

2. Useful (q,M) problems

The starting tableau in Lemke's algorithm for solving (q,M) is

$$\begin{array}{|c|c|c|c|}
\hline
 & x_o & x & r \\
\hline
q & -e & -M & I \\
\hline
\end{array}$$

where e is a vector of 1's, I is a diagonal matrix of 1's, the vector $r^T = (r_1, r_2, \ldots, r_n)$ that is eventually expected to satisfy $-Mx+Ir=q$ is in the basis, and x_o is a scalar artificial variable added to get started. The first pivot is defined by

x_o enters, r_1 leaves with 1 satisfying $q_1 = \min\{q_i : i \in \underline{n}\} < 0$

where $\underline{n} = \{1, 2, \ldots, n\}$ and q_1 is to be negative to avoid trivialities.

Theorem 1

Let $M \in R^{n \times n}$. Suppose

(1) $\qquad\qquad\qquad\qquad (M)_{ii} \neq 0 \quad \text{some } i \in \underline{n},$

(2) $\qquad\qquad\qquad\qquad (M)_{js} \neq 0 \quad \text{some } j, s \in \underline{n}, \; j \neq s.$

Then by (possibly) interchanging some entries of two appropriate rows of M and changing the sign of some entries of M, one can transform M into a matrix $\tilde{M}$ for which there is a vector q such that $(q, \tilde{M})$ has a complementary solution and it is not solvable by Lemke's algorithm.

Proof.

Fix $i \in \underline{n}$ so that (1) holds; let $\sum\limits_{j \neq k}$ stand for $\sum\limits_{j=1, j \neq k}^{n}$ and fix $s \in \underline{n}$ so that

(3) $\qquad\qquad \sum\limits_{j \neq s} |(M)_{sj}| \geq \max\left\{ \sum\limits_{j \neq k} |(M)_{kj}| : k \in \underline{n}, \; k \neq s \right\};$

thus row s is a row with the largest sum of the absolute values of its off diagonal entries.

By (2) one has $\sum_{j \neq s} |(M)_{sj}| > 0$, hence by (possibly) changing the sign of some entries of M one obtains a matrix M' satisfying $|(M')_{sj}| = |(M)_{sj}|$ all s, j, and

$$(M')_{ii} < 0, \qquad \sum_{j \neq s} (M')_{sj} > 0,$$

$$\sum_{j \neq s} (M')_{sj} \geq \varepsilon + \sum_{j \neq k} (M')_{kj}, \quad \text{all } k \in \underline{n}, \ k \neq s,$$

some $\varepsilon > 0$.

If i=s, take $\widetilde{M}$ to be M'. If i≠s interchange the off diagonal entries of the rows s and i of M' leaving $(M')_{ii}$ and $(M')_{ss}$ as they are, and call $\widetilde{M}$ the matrix so obtained; in detail for i < s such interchange defines the ith row of $\widetilde{M}$ as

$$(\widetilde{M})_{i1} = (M')_{s1}, \ldots, (\widetilde{M})_{i,i-1} = (M')_{s,s-1}, (\widetilde{M})_{ii} = (M')_{ii}, (\widetilde{M})_{i,i+1} = (M')_{si} \ ,$$

$$\ldots, (\widetilde{M})_{is} = (M')_{s,s-1}, (\widetilde{M})_{i,s+1} = (M')_{s,s+1}, \ldots, (\widetilde{M})_{in} = (M')_{sn}.$$

Thus, by changing some more signs if needed, the resulting matrix M can be taken to satisfy

(4) $\qquad (\widetilde{M})_{ii} < 0, \qquad (\widetilde{M})_{ii} \leq \min \left\{ (\widetilde{M})_{ji} : \ j \in \underline{n}, \ j \neq i \right\},$

(5) $\qquad \sum_{j \neq i} (\widetilde{M})_{ij} > 0, \qquad \sum_{j \neq i} (\widetilde{M})_{ij} \geq \varepsilon + \sum_{j \neq k} (\widetilde{M})_{kj} \quad \text{all } k \neq i$

some $\varepsilon > 0$. Let us now define $q \in R^n$ by

(6) $\qquad \begin{cases} q_k = - \sum_{j \neq k} (\widetilde{M})_{kj} & \text{for } k \in \underline{n}, \ k \neq i \\[2ex] q_i = \varepsilon - \sum_{j \neq i} (\widetilde{M})_{ij} \end{cases}$

where, since $\sum_{j \neq i} (\widetilde{M})_{ij} > 0$, $\varepsilon > 0$ is possibly reduced so that $q_i < 0$. It is to be shown that such $(q, \widetilde{M})$ has a complementary solution. Put

$$\bar{x}_k = \begin{cases} 1 & \text{for } k \in \underline{n}, \; k \ne i \\ 0 & \text{else,} \end{cases}$$

$$\bar{r}_k = \begin{cases} 0 & \text{for } k \in \underline{n}, \; k \ne i \\ \varepsilon & \text{else.} \end{cases}$$

One has $\bar{x} \ge 0$, $\bar{r} \ge 0$, $\bar{x}^T \bar{r} = 0$ and

$$(\widetilde{M}\bar{x})_k + q_k = 0 = \bar{r}_k \quad \text{all } k \in \underline{n}, \; k \ne i$$

$$(\widetilde{M}\bar{x})_i + q_i = \varepsilon = \bar{r}_i \, ,$$

i.e. $\widetilde{M}\bar{x} + q = \bar{r}$; thus $\bar{x}$ is a solution for $(q, \widetilde{M})$. It remains to be shown that $(q, \widetilde{M})$ cannot be solved by Lemke's algorithm. To see this let us observe that

(7) $\qquad\qquad q_i = \min \{ q_k : \; k \in \underline{n} \} < 0, \text{ and } q_i \ne q_k \text{ all } k \ne i;$

in fact for $k \in \underline{n}$, $k \ne i$ (5) says that

$$\varepsilon + \sum_{j \ne k} (\widetilde{M})_{kj} < \sum_{j \ne i} (\widetilde{M})_{ij} \, ,$$

thus in view of the definition of q

$$q_i = - \sum_{j \ne i} (\widetilde{M})_{ij} + \varepsilon < - \sum_{j \ne k} (\widetilde{M})_{kj} = q_k$$

and moreover $q_i < 0$ for the choice of ε. Hence (7) holds. To conclude the proof it is enough to show that Lemke's algorithm goes into a ray termination after one pivot. In fact, in view of (7) the first pivot is

entering variable x_o, leaving variable r_i,

hence by the complementary pivot rule the next entering variable is x_i; however the column of x_i after executing the pivot above is

$$-(\widetilde{M})_{ki} + (\widetilde{M})_{ii} \quad \text{for } k \in \underline{n}, \; k \neq i$$

$$(\widetilde{M})_{ii}$$

hence by (4) it has no positive entry. This shows that Lemke's algorithm has a ray termination and concludes the proof.

3. Example

Let the matrix

$$
M = \begin{bmatrix}
0 & 2 & 7 & -4 & 3 & 1 & 7 \\
2 & 7 & -6 & 0 & 0 & -4 & 1 \\
8 & 7 & 0 & -5 & 7 & 2 & 2 \\
5 & 1 & 7 & 0 & 4 & -5 & 7 \\
3 & 7 & -6 & 7 & 0 & 5 & 9 \\
6 & 4 & 1 & 7 & 6 & 0 & 2 \\
2 & 5 & 7 & 4 & -5 & 5 & 0
\end{bmatrix}
$$

be given. The following $(q, \widetilde{M})$ problem

$$
\widetilde{M} = \begin{bmatrix}
0 & 2 & 7 & -4 & 3 & 1 & 7 \\
3 & -7 & 7 & -6 & 7 & 5 & 9 \\
8 & 7 & 0 & -5 & 7 & 2 & 2 \\
5 & 1 & 7 & 0 & 4 & -5 & 7 \\
2 & -6 & 0 & 0 & 0 & -4 & 1 \\
6 & 4 & 1 & 7 & 6 & 0 & 2 \\
2 & 5 & 7 & 4 & -5 & 5 & 0
\end{bmatrix}
\quad , \quad
q = - \begin{bmatrix}
14 \\
25 \\
14 \\
18 \\
-1 \\
22 \\
14
\end{bmatrix}
$$

is obtained by making use of theorem 1.

References

[1] R. Chandrasekaran, "A special case of the complementary pivot problem" Opsearch 7 (1970) 263-268.

[2] M. Cirinà, "Complementarity", Atti Convegno AMASES (Perugia) 1981, 93-100.

[3] M. Cirinà, "Some (q,M) problems", Atti Giornate AIRO (Como) 1982, 289-297.

[4] R.W. Cottle, "Completely-$\mathcal{Q}$ matrices", Math. Progr. 19 (1980) 347-351.

[5] R.W. Cottle and G.B. Dantzig, "Complementary pivot theory of mathematical programming", Lin. Alg. and its Appl. 1 (1968) 103-125.

[6] B.C. Eaves, "The linear complementarity problem", Man. Science 17 (1971) 612-634.

[7] H.W. Kuhn and A.W. Tucker, "Nonlinear programming", in Proc. Second Berkeley Symposium on Mathematical Statistics and Probability, J. Neyman (editor), University of California Press (1951) 481-492.

[8] C.E. Lemke, "Bimatrix equilibrium points and mathematical programming" Man. Science 11 (1965) 681-689.

[9] O.L. Mangasarian, "Characterization of linear complementarity problems as linear programs", Math. Progr. Study 7 (1978) 74-87.

[10] K.G. Murty, "Computational complexity of complementary pivot methods" Math. Progr. Study 7 (1978) 61-73.

[11] A. Prekopa, "On the development of optimization theory", Amer. Math. Monthly 87 (1980) 527-542.

[12] Van Der Heyden, "A variable dimension algorithm for the linear complementarity problem", Math. Progr. 19 (1980) 328-346.

ASYMPTOTICALLY STABLE SOLUTIONS TO STOCHASTIC OPTIMIZATION PROBLEMS

Sjur D. Flåm
Chr. Michelsen Institute
N-5036 Fantoft NORWAY

ABSTRACT

The paper is concerned with a concave, infinite horizon, discrete time, stochastic optimization model. We characterize optimal solutions in terms of dual variables and prove that under appropriate conditions of strict concavity, all optimal trajectories will approach each other in distribution irrespective of starting point.

1.   INTRODUCTION

We consider the following <u>stochastic optimization problem</u>:

(P)   For a given initial point x_0 maximize $E \sum\limits_{t=0}^{\infty} u_t(\omega, x_t, x_{t+1})$
subject to the constraint that the decision x_t at stage $t \geq 1$ should be <u>nonanticipative</u>.

This constraint means that x_t should depend only on information available at time t.   More formally for $t \geq 1$, x_t is required to be F_t-<u>measurable</u> where $F_1 \subseteq F_2 \subseteq \ldots \subseteq F$ is a nest of σ-fields in some probability space (Ω, F, P).

For notational convenience we write $x_t \in F_t$ to denote that x_t is F_t-measurable.   We also write the <u>conditional expectation</u> $E(\cdot \mid F_t)$ briefly as E_t.

In (P), x_t belongs to R^n for all t, ω denotes the generic random outcome in Ω and $u_t : R^{2n} \to [-\infty, \infty)$ is the utility function in period t.

Let L_{2n}^{∞} be the space of all <u>essentially bounded</u> functions

$$y : \{0, 1, 2, \ldots \} \times \Omega \to R^{2n}.$$

Denote by N the space of all $y \in L_{2n}^{\infty}$ such that for each $t \geq 0$, $y_t = (z_t, w_t)$ where $z_t \in R^n$ is F_t-measurable and $w_t = z_{t+1}$.   Note that N is <u>closed</u> when L_{2n}^{∞} is endowed with the essential supremum norm.

We agree in (P) to consider only trajectories which are essentially bounded. Moreover, for $y = (z, w) \in L_{2n}^{\infty}$ the series

$$(1) \qquad \sum_{t=0}^{\infty} u_t(\omega, z_t, w_t)$$

is supposed to be majorized almost surely from above by at least one integrable function. For this reason we adopt the following convention: If (1) is not integrable, then its expectation equals $-\infty$. In order to exclude intractable cases we will assume that the supremum in (P) is less than $+\infty$.

It is essential to appreciate that all constraints in (P), except the insistence upon nonanticipativity, have been incorporated into the criterion by means of the value $-\infty$ expressing complete dissatisfaction. We emphasize that (P), being a discrete time stochastic version of the classical calculus of variations problem is generic for models of optimal resource management. Such models require an infinite planning horizon in order to be philosophically attractive.

The rest of the paper is organized as follows. Section 2 provides a characterization of optimality in terms of efficiency prices. We state in particular that if optimal solutions are strictly interior, then the dual solution constitutes an integrable function. In the case in which future utility is discounted, this means that the flow of current efficiency prices has a finite expected present value. Section 3 exploits this characterization to prove that in the strictly concave case optimal trajectories starting at different initial points will approach each other in distribution.

This result is the main contribution of this paper and it does not rely on uncertainty being represented in terms of a stationary process. Section 4 concludes the paper with some bibliographical remarks.

2. CHARACTERIZING OPTIMALITY

The purpose of this section is to demonstrate that a trajectory is optimal if and only if it is supported by an integrable regime of efficiency prices. The next section will exploit this result to prove asymptotic stability. To obtain the desired characterization of optimality we have to impose strong assumptions about concavity and strict feasibility.

Hereafter, suppose that u_t is a proper concave normal integrand on R^{2n} for each $t \geq 0$ (Rockafellar, 1976).

Theorem 2. Suppose that any optimal trajectory $(x_t)_{t=0}^{\infty}$ is strictly feasible in the following sense:

For some $\varepsilon > 0$

$$E\{u_0(\omega, x_0, x_1 + b) + \sum_{t=1}^{\infty} u_t(\omega, x_t + a, x_{t+1} + b)\} > -\infty$$

for all a, b ε R^n with $|a|$, $|b| < \varepsilon$. Then $(x_t)_{t=0}^{\infty}$ is optimal among all solutions of (P) starting at x_0 if and only if for some sequence of efficency prices $p_t : \Omega \to R^n$, $t = 1, 2, \ldots$, the following conditions are satisfied:

(i) P_t is F_t-measurable for all $t \geq 1$.

(ii) $E_0\{u_u(\omega, x_0, x_1') + p_1 x_1'\}$ is a.s. maximal over all $x_1' \varepsilon F_1$ at

$x_1' = x_1$, and for $t \geq 1$,

$E_t\{u_t(\omega, x_t', x_{t+1}') - p_t x_t' + p_{t+1} x_{t+1}'\}$

is a.s. maximal over all

$x_t' \varepsilon F_t$, $x_{t+1}' \varepsilon F_{t+1}$ at $x_t' = x_t$, $x_{t+1}' = x_{t+1}$.

(iii) $E \sum_{t=1}^{\infty} |p_t| < \infty$.

Proof. Suppose (i)-(iii) are all satisfied. Let $(x_t')_{t=0}^{\infty}$ be any feasible trajectory starting at x_0. Then for an abitrary $P_0 \varepsilon R^n$ and integer $T \geq 0$,

$$E \sum_{t=0}^{T} \{u_t(\omega, x_t, x_{t+1}) - u_t(\omega, x_t', x_{t+1}')\}$$

$$= E \sum_{t=0}^{T} E_t\{u_t(\omega, x_t, x_{t+1}) - u_t(\omega, x_t', x_{t+1}')\}$$

$$\geq E \sum_{t=0}^{T} E_t\{p_t(x_t - x_t') - p_{t+1}(x_{t+1} - x_{t+1}')\}$$

$$= E \sum_{t=0}^{T} \{p_t(x_t - x_t') - p_{t+1}(x_{t+1} - x_{t+1}')\}$$

$$= E \, p_{T+1}(x_{T+1}' - x_{T+1}) \to 0 \text{ as } T \to \infty.$$

Hence $(x_t)_{t=0}^{\infty}$ is optimal. Note that we did not need p_t to be F_t-measurable. Moreover, a transversality condition which is weaker than (iii) would suffice, see Weitzman, (1973).

We now turn to the necessity of (i)-(iii). Define the two functions ψ, δ : $L_{2n}^{\infty} \to [-\infty, \infty)$ by

$$\psi(y) = E\{u_0(\omega, x_0, w_0) + \sum_{t=1}^{\infty} u_t(\omega, z_t, w_t)\},$$

where $y_t := (z_t, w_t)$, and

$$\delta(y) = \begin{cases} 0 \text{ if } y_t = (z_t, w_t) \text{ with } z_t \in F_t \text{ and } w_t = z_{t+1} \\ -\infty \text{ otherwise.} \end{cases}$$

Suppose $(x_t)_{t=0}^{\infty}$ is optimal among all trajectories starting at x_0. For notational convenience denote $(x_t, x_{t+1})_{t=0}^{\infty}$ simply by x. The strict feasibility of x guarantees, by Rockafellar (1976), Corollary 2c, that the set $\partial\psi(x)$ of supergradients of ψ at x is a non-empty, weakly compact subset in the space of integrable functions π : $\{0, 1, 2, ...\} \times \Omega \to R^{2n}$. Note that (P) may be rephrased as:

maximize $(\psi + \delta)$ over L_{2n}^{∞}.

Since x is optimal we know that

$$0 \in \partial(\psi + \delta)(x).$$

By Ekeland and Temam (1976), Proposition I. 5.6, we may find a continuous linear functional π : $L_{2n}^{\infty} \to R$ such that

(1) $\pi \in \partial\psi(x)$ and

(2) $-\pi \in \partial\delta(x)$

The integrability of supergradients of ψ at x implies that

$$\pi = (\pi_{1t}, \pi_{2t})_{t=0}^{\infty}$$

where π_1, π_2 : $\{0, 1, 2, ...\} \times \Omega \to R^n$ are both integrable.

By (1) we have for every $t \geq 1$,

(3) $u_t(\omega, z_t, w_t) - \pi_{1t} z_t - \pi_{2t} w_t$ is maximal a.s. at
        $z_t = x_t$, $w_t = x_{t+1}$, and in addition

(4) $u_0(\omega, x_0, w_0) - \pi_{20} w_0$ is maximal a.s. at $w_0 = x_1$.

Take the conditional expectation E_t in (3) to obtain that

(5) $E_t\{u_t(\omega, x_t', , x_{t+1}') - q_t x_t' + p_{t+1} x_{t+1}'\}$

is almost surely maximal over all $x_t' \in F_t$, $x_{t+1}' \in F_{t+1}$ at

$x_t' = x_t$, $x_{t+1}' = x_{t+1}$.

Here $p_{t+1} = -E_{t+1}\pi_{2t}$ and $q_t = E_t\pi_{1t}$ for all $t \geq 0$. We note that (2) implies that π is orthogonal to N. For an arbitrary $T \geq 1$, let $(x_t')_{t=0}^{\infty}$ be such that $x_t' = x_t$ for all $t \neq T$ and $x_T' = x_T + a$ where $a \in F_T$

(2) tells that P is orthogonal to N at x. Thus

$$0 = \pi(x'-x) = E \sum_{t=0}^{\infty} \{\pi_{1t}(x_t' - x_t) + \pi_{2t}(x_{t+1}' - x_{t+1})\}$$

$$= E(\pi_{1T} + \pi_{2,T-1})a$$

Letting a be any F_T-measurable function we see that $E_T(\pi_{1T} + \pi_{2,T-1}) = 0$ for all $T \geq 1$.
It follows that $q_T = p_T$ for all $T \geq 1$.

Taking the conditional expectation E_1 in (4) and thereafter E_0 we obtain that

$E_0\{u_0(\omega, x_0, x_1') + p_1 x_1'\}$

is maximal over all $x_1' \in F_1$ at $x_1' = x_1$. This completes the proof.   Q.E.D.

3. ASYMPTOTIC STABILITY OF OPTIMAL SOLUTIONS

Consider two different optimal solutions $(x_t)_{t=0}^{\infty}$ and $(x_t')_{t=0}^{\infty}$ starting at different initial points. The purpose of this section is to demonstrate that x_t and x_t' approach each other as $t \to \infty$. Specifically we wish to show that $x_t - x_t' \to 0$

in _distribution_, which implies $x_t - x_t' \to 0$ in _probability_. See Billingsley (1968, Section 1.4).

To achieve this we have to impose stronger concavity assumptions.

Definition

The family u_t, $t \geq 0$ is said to be _uniformly concave_ if for all $\epsilon > 0$ there exists $\delta > 0$ such that $|y - y'| \geq \epsilon$ implies

$$u_t(\omega, \frac{y+y'}{2}) \geq \frac{1}{2} \{u_t(\omega, y) + u_t(\omega, y') + \delta\} \text{ a.s. for all } t \geq 0.$$

See Zalinescu (1983), Theorem 2.2 for characterizations of uniform concavity in terms of differentiation. We now state the chief result in this paper.

Theorem 2. Suppose any optimal solution to (P) is strictly feasible as described in Theorem 1. Also suppose that the family u_t, $t \geq 0$, is uniformly concave. Then for any two optimal trajectories $(x_t)_{t=0}^{\infty}$ and $(x_t')_{t=0}^{\infty}$ starting at different initial points we have that $x_t - x_t' \to 0$ in distribution.

Proof. Let $p = (p_t)_{t=1}^{\infty}$ and $p' = (p_t')_{t=1}^{\infty}$ be sequences of efficiency prices supporting the optimal trajectories $x = (x_t)_{t=0}^{\infty}$ and $x' = (x_t')_{t=0}^{\infty}$ respectively as described in Theorem 1. By Theorem 1, (ii), for $t \geq 1$,

(6) $$E_t(u_t(\omega, x_t, x_{t+1}) - p_t x_t + p_{t+1} x_{t+1}) \geq$$

$$E_t(u_t(\omega, x_t', x_{t+1}') - p_t x_t' + p_{t+1} x_{t+1}') \text{ and}$$

(7) $$E_t(u_t(\omega, x_t', x_{t+1}') - p_t' x_t' + p_{t+1}' x_{t+1}') \geq$$

$$E_t(u_t(\omega, x_t, x_{t+1}) - p_t' x_t + p_{t+1}' x_{t+1}).$$

Adding (6) and (7) we obtain

(8) $$E_t\{(p_{t+1} - p_{t+1}')(x_{t+1} - x_{t+1}') - (p_t - p_t')(x_t - x_t')\} \geq 0.$$

Define $v_t := (p_t - p_t')(x_t - x_t')$.

Then (8) reads

(9) $B_t(v_{t+1} - v_t) \geq 0.$

Taking the expectation in (9) we obtain

(10) $Bv_{t+1} \geq Bv_t$ for all $t \geq 1$.

The fact that p, p' are both integrable and x, x' are both essentially bounded implies by (10) that

(11) $Bv_t \to 0.$

Now suppose $x_t - x_t'$ does not converge in distribution to 0. Then for some ε, μ>0

(12) $P\{\omega: ||(x_t(\omega), x_{t+1}(\omega)) - (x_t'(\omega), x_{t+1}'(\omega))|| \geq \epsilon\} \geq$

$P\{\omega: ||x_t(\omega) - x_t'(\omega)|| \geq \epsilon\} \geq \mu$

for infinitely many t.

By Zalinescu (1983), Theorem 2.2, (IV) there exists a convex lower semi-continuous function

$\psi : [0, \infty) \to [0, \infty]$ with int dom $\psi \neq \emptyset$, $\psi(d) = 0$ iff $d = 0$

such that

(13) $\langle \bar{y} - \bar{y}', y - y' \rangle \leq -\psi(||y - y'||)$
        when $\bar{y} \in \partial u_t(\omega, y)$ and $\bar{y}' \in \partial u_t(\omega, y')$.

Suppose y, y' are both F_{t+1}-measurable. Taking the conditional expectation B_{t+1} in (13) we get

(14) $\langle B_{t+1}(\bar{y} - \bar{y}'), y - y' \rangle \leq -\psi(||y - y'||).$

Let $y = (x_t, x_{t+1})$, $\bar{y} = (\pi_{1t}, \pi_{2t})$ and $y' = (x_t', x_{t+1}')$, $\bar{y} = (\pi_{1t}', \pi_{2t}')$ as in the proof of Theorem 1. Then (13) and (14) imply

(15) $\quad (p_{t+1} - p'_{t+1})(x_{t+1} - x'_{t+1}) - \langle E_{t+1}(\pi_{1t} - \pi_{2t}), x_t - x'_t \rangle$

$\qquad \geq \psi(||(x_t, x_{t+1}) - (x'_t, x'_{t+1})||).$

Taking the expectation in (15) we arrive at

(16) $\quad E\{(p_{t+1} - p'_{t+1})(x_{t+1} - x'_{t+1}) - (p_t - p'_t)(x_t - x'_t)\}$

$\qquad \geq E\psi((||(x_t, x_{t+1}) - (x'_t, x'_{t+1})||)$

Now (12) and (16) imply that $E(v_{t+1} - v_t) \geq \mu\psi(\epsilon)$ for infinitely many t. This contradicts (11), however. Hence $x_t - x'_t \to 0$ in distribution.

$\qquad\qquad\qquad\qquad\qquad\qquad\qquad\qquad\qquad\qquad\qquad$ Q.E.D.

4. BIBLIOGRAPHICAL REMARKS

Since (P) is a problem of the classical calculus of variations type, optimality should be partially characterized by the Euler equation. In fact Theorem 1, (ii) is a discrete time stochastic version of this equation. Similar results have been obtained in the finite horizon Bolza (i.e. free end-point) problem by Rockafellar and Wets (1983). Zilcha (1976) has also given a characterization of optimality in terms of prices. However, he imposes assumptions about free disposal and "nothing ventured, nothing gained" (see also Weitzman, 1973).

The asymptotic convergence of optimal paths has been proved by Jeanjean (1974). However, he represents uncertainty in terms of a Markov chain (see also Donaldson and Mehra, 1983). With this specification, or even more generally with uncertainty being stationary, one may prove that an optimal steady state exists (see Jeanjean 1974, Evstigneev 1974, Majumdar and Radner 1983, Flåm 1983). This state is a common limiting distribution of all optimal trajectories.

Observe that $E(p_t - p'_t)(x_t - x'_t)$ served as a Liapunov function in the proof of Theorem 2. This dual approach to stability is carried over from deterministic models. See Cass and Shell (1976), or McKenzie (1976).

We conclude by remarking that uniform concavity is not satisfied when future utilities are discounted. In this case some curvature condition should be placed directly on the Hamiltonian of the system (Brock and Majumdar 1978, Flåm 1983).

REFERENCES

1. Billingsley, P.   Convergence of Probability Measures.
                                        J. Wiley, N.Y. (1968).

2. Brock, W.A. and Majumdar, M.   Global asymptotic stability results
                                        for multisector models of optimal
                                        growth under uncertainty when future
                                        utilities are discounted.
                                        J. of Economic Theory 18, (1978),
                                        225-243

3. Cass, D, and Shell, K.   The structure and stability of
                                        competitive dynamical systems.
                                        J. of Economic Theory 12, (1976),
                                        31-70.

4. Donaldson, J.B. and Mehra, R.   Stochastic growth with correlated
                                        production shocks.
                                        J. of Economic Theory 29, (1983),
                                        282-312.

5. Ekeland, I. and Temam, R.   Convex Analysis and Variational
                                        Problems.
                                        North Holland, N.Y. (1976).

6. Evstigneev, I.V.   Optimal stochastic programme and
                                        their stimulating prices. In: J.
                                        Los & W. Los, eds. Mathematical
                                        Models in Economics. N. Holland,
                                        Amsterdam (1974).

7. Flåm, S.D.   Turnpike results in stochastic models.
                                        Technical report 832155-6, CMI,
                                        Bergen (1983).

8. Jeanjean, P.   Optimal development programs under
                                        uncertainty.
                                        J. of Economic Theory 7, (1974),66-92.

9.   Majumdar, M. and Radner, R.   Stationary optimal policies
                                            withdiscounting in a stochastic
                                            activity analysis.
                                            Econometrica 51, 6, (1983) 1821-1837.

10.   McKenzie, L.W.   Turnpike theory.
                                            Econometrica 44, (1976), 841-865.

11.   McKenzie, L.W.   A primal route to the turnpike and
                                            Liapounov stability.
                                            J. of Economic Theory 27, 194-209
                                            (1982).

12.   Rockafellar, R.T.   Integral Functionals, Normal
                                            Integrands and Measurable Selections.
                                            Lecture Notes in Mathematics
                                            no. 543. Springer Verlag N.Y. (1976).

13.   Rockafellar, R.T. and Wets, R.J-B.   Deterministic and stochastic
                                            optimization problems of Bolza type
                                            in discrete time.
                                            Stochastics, Vol. No. 10.3.4, (1983).

14.   Weitzman, M.L.   Duality theory for infinite horizon
                                            convex models.
                                            Management Science 19, (1973),
                                            783-789.

15   Zalkinescu, C.   On uniformly convex functions.
                                            J. of Math. Anal. and Appl. 95,
                                            (1983), 344-374.

15.   Zilcha, I.   Characterization by prices of optimal
                                            programs under uncertainty.
                                            J. of Mathematical Economics 3,
                                            (1976), 173-183

ON INTEGRATED CHANCE CONSTRAINTS

W.K. Klein Haneveld

Institute of Econometrics
University of Groningen
P.O. Box 800, 9700 AV

§1. <u>Introduction</u>. In the literature on stochastic programming two different approaches are well-known: chance constrained programming (CCP) and stochastic programming with recourse (SPR). Many attention has been paid to the relation of both modeling techniques. Several authors ([2],[3],[11],[12], [14]) established certain equivalencies between CCP and SPR. The results are not completely convincing, however. For example, CCP problems may be non-convex whereas SPR problems are always convex ([5] p. 90). Moreover, mathematical equivalence is not to be interpreted as economic equivalence [2]): in applications the specification of a probability level of feasibility might be more appropriate than the specification of penalty costs for infeasibility, or just reverse, depending on the circumstances. Some authors ([4]) try to show, that CCP is inferior to SPR as a modeling technique, neglecting e.g. the possibility that one may combine both approaches ([8]). Also their conclusion is biased, since they do not pay the same critical detailed attention to the specification of recourse costs in practice as they do to the specification of reliability levels in CCP.

Both CCP and SPR models deal with the risks of infeasibility in linear programs with random coefficients. One difference between both approaches which does not get enough attention in the literature, is the fact that the risk is measured differently: CCP measures the risk of infeasibility *qualitatively* whereas SPR does it *quantitatively*. That is, in CCP the possibility of infeasibility is at stake regardless the amounts by which the constraints are violated, whereas in SPR these amounts are important. In practice it might be acceptable to have a constraint violated, even with high probability if necessary, if the amount by which it is violated is small enough. In particular, this might be the case if the constraint represents a quantitative goal to be reached rather than a logical or technical necessity, as is often the case in practical linear programming models. In such cases the quantitative approach of infeasibility of SPR is appropriate. However, this does not mean that one has to adopt the penalty cost (in [1] called discrepancy cost) structure of SPR at the same time. It is quite conceivable that in certain circumstances it is more appropriate to fix an upper bound on the risk (as in CCP) rather than to penalize the risk (as in SPR). This leads to constraints, where the mean value of the amount, by which the underlying constraint is violated, is bounded from above. In this paper we analyze several variants of such *integrated chance constraints* (ICCs); the name will be explained in §2.

In §2 two different ICC formulations for single random constraints are studied. In the first type there is a fixed upperbound on the risk, and in the second one the upperbound depends on the functions and distributions involved. Both give rise to convex feasibility sets, which increase strictly and continuously with the risk aversion parameter. In §3 ICC formulations are given for the joint risk of infeasibility of a system of random inequalities. It appears that the analogue of the first type of single ICC has the same nice behaviour. In each of the three cases there is an intimate relation with SPR models, as shown in §4. The conclusions are summarized in §5.

§2. Single Integrated Chance Constraints. We consider a linear programming model with random coefficients in the constraints, assuming that these coefficients are only known in distribution. In particular, we are interested in the question under which condition a decision vector x is called feasible. In this section we consider one scalar constraint in $x \in \mathbb{R}^n$,

(1) $\Sigma_{j=1}^n a_j x_j \geq b$,

where the vector $(a_1, \ldots, a_n, b)$ is random, with known probability distribution, of which the mean values $(\bar{a}_1, \ldots, \bar{a}_n, \bar{b})$ are finite. Defining $\eta(x) := \Sigma_{j=1}^n a_j x_j - b$, $[\eta(x)]^- := \max(0, -\eta(x))$, we shall call the random variable $[\eta(x)]^-$ the *shortage* determined by (1). The constraint (1) reflects the idea that one wants to avoid positive shortage, but because of the random character it is impossible in general or undesirable to exclude shortages completely. Depending on the *definition of risk* together with the specification of the maximal risk level (i.e. the *degree of risk aversion*) one gets different *risk constraints* as a translation of (1). In the most well-known case risk is understood as *probability of positive shortage*, resulting in a feasible set determined by a *chance constraint* (CC)

(2) $X_0(\alpha) := \{x \in \mathbb{R}^n : \text{Esgn}[\eta(x)]^- \leq \alpha\}, \alpha \in [0,1]$.

Notice that $\text{Esgn}[\eta(x)]^- = P(\eta(x) < 0)$ so that $x \in X_0(\alpha)$ iff $P(\Sigma_{j=1}^n a_j x_j \geq b) \geq 1-\alpha$. The risk aversion parameter α denotes the maximal acceptable risk level. We like to stress, that for this specification of risk the *amount of shortage* is not relevant; only its sign counts. In situations where this amount is important, it might be more appropriate to use the *mean shortage* $E[\eta(x)]^-$ as a measure for risk. This leads to the following analogue of (2)

(3) $X_1(\beta) := \{x \in \mathbb{R}^n : E[\eta(x)]^- \leqq \beta\}, \beta \in [0,\infty).$

Since

(4) $E[\eta(x)]^- = \int_{-\infty}^0 P(\eta(x) < t)dt$

we shall call (3), and other constraints asking for upperbounds on $E[\eta(x)]^-$, *integrated chance constraints (ICC)*. Formally, one might introduce (3) by starting with chance constraints for each $t \in (-\infty,0]$

(5) $P(\eta(x) < t) \leq \alpha_t , \alpha_t \in [0,1]$

where $\alpha_t \downarrow 0$ for $t \downarrow -\infty$, and replacing (5) by the integrated version (3), with $\beta = \int_{-\infty}^0 \alpha_t dt$. In (3) the risk aversion parameter β is fixed, and supposed to be choosen a priori. Whereas α in (2) is scale-free, β in (3) depends on scale. If the "demand" b is fixed and positive, one may choose e.g. $\beta = \alpha b$ for a scale-free $\alpha \in [0,1]$. It is also possible to specify the maximal accepted risk not as a fixed number β, but depending on the distribution of $\eta(x)$. For example, since a natural upperbound for $E[\eta(x)]^-$ is given by $E|\eta(x)|$, we introduce

(6) $X_2(\alpha) := \{x \in \mathbb{R}^n: E[\eta(x)]^- \leqq \alpha E|\eta(x)|\}, \alpha \in [0,1].$

In (6) also the mean surplus $E[\eta(x)]^+ = E\max(0,\eta(x))$ is taken into account, since $E[\eta(x)]^- + E[\eta(x)]^+ = E|\eta(x)|$. Just like (3) also (6) is a direct analogue of (1), as can be seen by rewriting the CC as: $E\text{sgn}[\eta(x)]^- \leq \alpha E\text{sgn}|\eta(x)|$ assuming that $P(\eta(x) = 0) = 0$. In addition to (3) and (6) we call the attention to the integrated chance constraint of the type

(7) $X_3(\gamma) := \{x \in \mathbb{R}^n : E[\eta(x)]^- \leqq \gamma \cdot P(\eta(x) < 0)\}, \gamma \in (0,\infty),$

analyzed by Prékopa ([8]). Here γ is the maximal accepted value for the conditional mean shortage $E[[\eta(x)]^-|\eta(x) < 0]$.

In the sequel of this section we shall analyze the mathematical properties of the ICCs (3) and (6). Roughly speaking, we shall show that they have important properties which CCs (and also (7)) do not have, at least not in general. Moreover, in the cases for which easy deterministic equivalent formulations for $X_0(\alpha)$ exist,   the same is true for $X_1(\beta)$ and $X_2(\alpha)$. It is not surprising that the ICC formulations behave better than their CC companions, since the function $\psi(z):= [z]^-$ is continuous and strictly decreasing for $z < 0$, which is not true for the function sgn $\psi(z)$. Throughout this paper we assume

(8) $E|a_j| < \infty$, $j = 1,\ldots,n$, $E|b| < \infty$.

Lemma 1. (a) The function $f(x):= E[\eta(x)]^-$ is nonnegative, finite, convex and Lipschitz continuous for all distributions of (a,b) satisfying (8). For finite discrete distributions the function f is piecewise linear. If the distribution of (a,b) has a density function, then f has a continuous gradient, with

(9) $\dfrac{\partial f}{\partial x_j}(x) = E[-a_j \cdot \text{sgn}[\eta(x)]^-]$.

(b) Moreover,

(10) $\lim_{\lambda \to \infty} \dfrac{f(x+\lambda y)-f(x)}{\lambda} = E[\sum_{j=1}^n a_j y_j]^-$, $y \in \mathbb{R}^n$.

Proof. (a) Although a direct proof is not difficult, we may simply refer to more general results on SPR ([5],[6]) since

$$f(x) = E[\min_{y \in \mathbb{R}^2} \{y_1 : y_1 - y_2 = b - \Sigma_j a_j x_j; \ y_1, y_2 \geqq 0\}]$$

can be seen as the second stage of a simple recourse model. (b) For any pair (p,q) of real numbers, satisfying $p = 0$ or $|p| \geq |q|$ the following equality holds:

(11) $[p+q]^- = [p]^- + [q]^- - \{[q]^- \cdot \text{sgn}[p]^+ + [q]^+ \cdot \text{sgn}[p]^-\}$.

By the substitution $p:= \lambda\Sigma_{j=1}^n a_j y_j$ and $q:= \eta(x)$ one derives from (11) that for any fixed $(a,b) \in \mathbb{R}^{n+1}$

$$\lim_{\lambda \to \infty}\lambda^{-1}\{[\eta(x+\lambda y)]^- - [\eta(x)]^-\} = [\Sigma_{j=1}^n a_j y_j]^-.$$

Moreover, $|\lambda^{-1}\{[\eta(x+\lambda y)]^- - [\eta(x)]^-\}| \leqq |\Sigma_{j=1}^n a_j y_j|$ for any $\lambda \neq 0$, and this majorant is integrable because of (8). Hence, the bounded convergence theorem of Lebesgue gives (10).   □

Since the $X_1(\beta)$, $\beta \geqq 0$, are level sets of $f(x) = E[\eta(x)]^-$, we have

Theorem 2. (Characterization of $X_1(\beta)$, $0 \leqq \beta < \infty$.) Assume (8). (a) $X_1(\beta)$ is a closed $convex$ subset of $\mathbb{R}^n$, polyhedral if (a,b) has a discrete distribution. Define

(12) $\beta_o := \inf\limits_{x \in \mathbb{R}^n} E[\eta(x)]^-.$

Then $X_1(\beta) \neq \emptyset$ iff $\beta > \beta_0$ or $\beta = \beta_0$ and the infimum in (12) is attained. Also $X_1(\beta) \neq \mathbb{R}^n$ unless $P(a=0) = 1$ and $\beta \geq E[b]^+$.

(b) $X_1(\beta)$ is nondecreasing with β. $X_1(0) = \{x \in \mathbb{R}^n : P(\eta(x) < 0) = 0\}$ and $X_1(\infty) := \cup\{X_1(\beta) : 0 \leq \beta < \infty\} = \mathbb{R}^n$. The increase is strict, generally; that is:

(13)   if $\emptyset \neq X_1(\beta) \neq \mathbb{R}^n$ then $X_1(\beta') \neq X_1(\beta)$ for all $\beta' \neq \beta$.

The increase is also continuous, generally; that is

(14) $X_1(\beta) = \cap_{\beta' > \beta} X_1(\beta')$ for all $\beta \geq 0$,

and

(15) $X_1(\beta) = cl \cup_{\beta' < \beta} X_1(\beta')$ for all $\beta > 0$

unless $\beta_0 > 0$, $\beta = \beta_0$ and $X_1(\beta_0) \neq \emptyset$.

(c) Let $X_1(\beta) \neq \emptyset$. Then $x + \lambda y \in X_1(\beta)$ for all $x \in X_1(\beta)$ and for all $\lambda \geq 0$ iff $P(\Sigma_{j=1}^n a_j y_j < 0) = 0$. Therefore, $X_1(\beta)$ is bounded iff $0 \in int\ conv\ S_a$, where $S_a \subset \mathbb{R}^n$ is the support of the distribution of the random vector $a = (a_1, \ldots, a_n)$.

Proof. (a) Follows from lemma 1. The last statement is a consequence of the fact that a bounded convex function is constant. (b) $E[\eta(x)]^- = 0$ iff $P(\eta(x) < 0) = 0$. $X_1(\infty) = \mathbb{R}^n$ since $E[\eta(x)]^-$ is finite for all $x \in \mathbb{R}^n$. In order to prove (13), assume $\emptyset \neq X_1(\beta) \neq \mathbb{R}^n$, so that $x_i \in \mathbb{R}^n$, $i = 1,2$, exist with $f(x_1) \leq \beta$ and $f(x_2) > \beta$. Since f is continuous, each value in $[f(x_1), f(x_2)]$ is attained, so that for any $\beta' > \beta$ an $\hat{x} \in \mathbb{R}^n$ exists with $\beta < f(\hat{x}) \leq \beta'$. Hence $X_1(\beta') \neq X_1(\beta)$. For $\beta' < \beta$ there are two possibilities: either $X_1(\beta') = \emptyset$ and we have nothing to prove, or $\exists x_3 \in \mathbb{R}^n$ with $f(x_3) \leq \beta'$. In the latter case one proves $X_1(\beta') \neq X_1(\beta)$ similarly to the case $\beta' > \beta$. (14) is a trivial property of level sets, and (15) is a direct consequence of (the finiteness and) the convexity of f ([10] p. 59). (c) Since f is finite and convex, all nonempty level sets $X_1(\beta)$ have the same recession cone, namely $C = \{y \in \mathbb{R}^n : \lim_{\lambda \to \infty} \lambda^{-1}(f(x+\lambda y)-f(x)) \leq 0\}$ where $x \in \mathbb{R}^n$ is arbitrary ([10] Ths 8.7 and 8.5). Therefore, the first statement in (c) follows from (10) since $E[\Sigma_{j=1}^n a_j y_j]^- \leq 0$ iff $P(\Sigma_{j=1}^n a_j y_j < 0) = 0$. Also, $X_1(\beta)$ is bounded iff $C = \{0\}$ ([10] Th. 8.4); equivalently iff $P(\Sigma_{j=1}^n a_j y_j \geq 0) < 1$ for all $y \neq 0$. This is precisely the case if $\{0\}$ can not be separated from the support S_a by a hyperplane, and that is true iff 0 is an interior point of S_a.   □

We shall now analyze the second type of ICC, introduced in (6). Since $z = [z]^+ - [z]^-$, $|z| = [z]^+ + [z]^-$, we have

(16a) $X_2(\alpha) = \{x \in \mathbb{R}^n : E[\eta(x)]^- \leq \alpha \, E|\eta(x)|\}$

(16b) $= \{x \in \mathbb{R}^n : (1-\alpha)E[\eta(x)]^- \leq E[\eta(x)]^+\}$

(16c) $= \{x \in \mathbb{R}^n : (1-2\alpha)E[\eta(x)]^- \leq \alpha E\eta(x)\}$.

It shows e.g. that constraints of the type (6) imply, that positive mean shortage is only accepted if the corresponding mean surplus is large enough. Excluding an uninteresting case, we shall assume

(17) $P(a=0) < 1$

so that $M := \{x \in \mathbb{R}^n : E|\eta(x)| > 0\} \neq \emptyset$, and we define

(18) $\alpha_0 := \inf_{x \in M} E[\eta(x)]^- / E|\eta(x)|$

$\alpha_1 := \sup_{x \in M} E[\eta(x)]^- / E|\eta(x)|$

   Theorem 3. *(Characterization of* $X_2(\alpha)$, $0 \leq \alpha \leq 1$.) Assume (8) and (17).
(a) $X_2(\alpha)$ is a closed subset of $\mathbb{R}^n$. For $\alpha = \frac{1}{2}$, it is the *linear* halfspace

(19) $X_2(\frac{1}{2}) = \{x \in \mathbb{R}^n : \sum_{j=1}^n \bar{a}_j x_j \geq \bar{b}\}$.

If $\alpha \leq \frac{1}{2}$ $X_2(\alpha)$ is *convex*, even polyhedral if (a,b) has a finite discrete distribution. If $\alpha > \frac{1}{2}$ $X_2(\alpha)$ is the complement of a convex set, hence nonconvex generally.
$X_2(\alpha) = \emptyset$ iff $M = \mathbb{R}^n$ and either $\alpha < \alpha_0$ or $\alpha = \alpha_0$ and the infimum in (18) is not attained. $X_2(\alpha) = \mathbb{R}^n$ iff $\alpha \geq \alpha_1$.
(b) $X_2(\alpha)$ is nondecreasing with α. $X_2(0) = \{x \in \mathbb{R}^n : P(\eta(x) < 0) = 0\}$,
$X_2(1^-) := \bigcup_{\alpha<1} X_2(\alpha) = \{x \in \mathbb{R}^n : P(\eta(x) \leq 0) < 1 \text{ or } P(\eta(x) = 0) = 1\}$ and
$X_2(1) = \mathbb{R}^n$. If M is connected the increase is *strict,* generally, that is

(20)   if $X_2(0) \neq X_2(\alpha) \neq \mathbb{R}^n$ then $X_2(\alpha') \neq X_2(\alpha)$ for all $\alpha' \neq \alpha$.

The increase is also *continuous,* generally, that is

(21) $X_2(\alpha) = \bigcap_{\alpha'>\alpha} X_2(\alpha')$ for all $\alpha \in [0,1)$

and

(22) $X_2(\alpha) = \text{cl} \bigcup_{\alpha'<\alpha} X_2(\alpha')$ for all $\alpha \in (0,1]$, $\alpha \neq \alpha_0$, $\alpha \neq \alpha_1$.

(c) For $\alpha \in [\frac{1}{2}, 1]$ $X_2(\alpha)$ is nonempty and unbounded. If $\alpha \in [0, \frac{1}{2}]$ and $X_2(\alpha) \neq \emptyset$ then $x + \lambda y \in X_2(\alpha)$ for all $x \in X_2(\alpha)$ and for all $\lambda \geq 0$ iff $E[\sum_{j=1}^n a_j y_j]^- \leq \alpha E|\sum_{j=1}^n a_j y_j|$. Therefore, $X_2(\alpha)$ is bounded iff for all $y \neq 0$ $E[\sum_{j=1}^n a_j y_j]^- > \alpha E|\sum_{j=1}^n a_j y_j|$. This condition can only be true if $0 \in \text{int conv } S_a$, where $S_a \subset \mathbb{R}^n$ is the support of the distribution of the random vector $a = (a_1, \ldots, a_n)$.

Proof. (a) Because of (16) we may write $X_2(\alpha) = \{x \in \mathbb{R}^n : h_\alpha(x) \leq 0\}$ where $h_\alpha(x) := (1-2\alpha)E[\eta(x)]^- - \alpha E\eta(x)$. Since $E\eta(x) = \sum_{j=1}^n \bar{a}_j x_j - \bar{b}$ it follows from lemma 1 that h is finite and continuous, convex if $\alpha \leq \frac{1}{2}$. If $x \notin M$ then $x \in X_2(\alpha) \forall \alpha$; therefore $X_2(\alpha) = \emptyset$ implies $M = \mathbb{R}^n$. (b) $X_2(0) = X_1(0)$. If $E|\eta(x)| = 0$ then $x \in X_2(\alpha)$ for all α; if $E|\eta(x)| > 0$ then $x \in X_2(\alpha)$ for an $\alpha < 1$ iff $E[\eta(x)]^- < E|\eta(x)|$, or $E[\eta(x)]^+ > 0$. This proves the formula for $X_2(1^-)$. If M is connected then $E[\eta(x)]^-/E|\eta(x)|$ is a continuous function on M, and the proof of (20) is similar to that of (13). (21) is a direct consequence of the continuity of $h_\alpha(x)$ as a function of α. Because of the monotonicity of $X_2(\alpha)$ in (22) only

(23) $\qquad X_2(\alpha) \subset \text{cl} \cup_{\alpha' < \alpha} X_2(\alpha')$

has to be proved. Suppose that (23) is not true. Then an $x_0 \in X_2(\alpha)$ exists and a neighbourhood $N(x_0)$ such that $N(x_0) \cap X_2(\alpha') = \emptyset \; \forall \alpha' < \alpha$. This implies $N(x_0) \subset M$ (since $\mathbb{R}^n \setminus M \subset X_2(\alpha') \forall \alpha'$) and $\alpha \geq \alpha_0$. Moreover, it implies that

(24) $\qquad h_\alpha(x) \geq 0 \; \forall x \in N(x_0)$, with equality for $x = x_0$,

so that x_0 is a local minimum of h_α. If $\alpha \leq \frac{1}{2}$ then the convexity of $h_\alpha(x)$ implies that the minimum is global, and therefore $h_\alpha(x) \geq 0 \; \forall x \in \mathbb{R}^n$. Consequently, $\alpha = \alpha_0$. If $\alpha > \frac{1}{2}$ h_α is concave, and a local minimum must be a global maximum, actually, so that $h_\alpha(x) = 0 \; \forall x \in \mathbb{R}^n$. Consequently, $\alpha \geq \alpha_1$; in fact, $\alpha = \alpha_1$ (otherwise $X_2(\alpha') = \mathbb{R}^n$ for α' sufficiently close to α, implying $N(x_0) \cap \cap X_2(\alpha') \neq \emptyset$). We conclude that (23) is true, except possibly if $\alpha = \alpha_0$ or $\alpha = \alpha_1$.

(c) Because of (8) and (17) $X_2(\frac{1}{2})$ is a nonempty halfspace, so that $X_2(\alpha)$ is nonempty and unbounded for $\alpha \geq \frac{1}{2}$. If $\alpha \leq \frac{1}{2}$ $X_2(\alpha)$ is a level set of the convex function h_α. From (10) it follows that $h_\alpha'(y) := \lim_{\lambda \to \infty} \lambda^{-1}$ · $(h_\alpha(x + \lambda y) - h_\alpha(x)) = (1-2\alpha)E[\sum_{j=1}^n a_j y_j]^- - \alpha E \sum_{j=1}^n a_j y_j = E[\sum_{j=1}^n a_j y_j]^- - \alpha E|\sum_{j=1}^n a_j y_j|$; hence, $\lambda > 0$ and $x \in X_2(\alpha)$ imply $x + \lambda y \in X_2(\alpha)$ iff this limit is nonpositive ([10] Thms 8.7 and 8.5). Therefore, in order that $X_2(\alpha)$ is bounded it is necessary that $\alpha < \frac{1}{2}$, and in that case the necessary and sufficient condition is: $h_\alpha'(y) > 0, \; \forall y \neq 0$. This condition is stronger than $E[\sum_{j=1}^n a_j y_j]^- > 0 \forall y \neq 0$, which is equivalent to $0 \in \text{int conv } S_a$ as shown in theorem 2. $\qquad \square$

The conditions for strict and continuous increase of $X_1(\beta)$ and $X_2(\alpha)$ are very weak. Under (18), the only possible exception occurs for $\beta = \beta_0$, and this is an extremely low specification of β (either $\beta_0 = 0$ or $\beta_0 > 0$ and $X_1(\beta) = \emptyset$ for all $\beta < \beta_0$). Similarly, in (22) α_0 and α_1 are extreme specifications of α(if $\alpha_0 > 0$ then $X_2(\alpha) = \{x: P(\eta(x) = 0) = 1\} = \mathbb{R}^n \setminus M$=: M^c for all $\alpha < \alpha_0$, and $X_2(\alpha) = \mathbb{R}^n$ for all $\alpha \geq \alpha_1$). If $\alpha_0 < \alpha < \alpha_1$ $X_2(\alpha)$ is strictly increasing (see (20)) if M is connected; and M can only be disconnected in very special cases. In order to see this, notice that $M^c = \{x: P(\Sigma_{j=1}^n a_j x_j = b) = 1\}$ is a linear variety in $\mathbb{R}^n$. M^c can separate M into disjoint parts only if it is a hyperplane, and in that case all a_j and b are linear functions of the same random variable. An illustration is given by the next two examples, where in both cases M^c consists of the origin only.

Example 1. n = 2, $P(b=0) = 1$, $P((a_1,a_2)=(-2,1)) = P((a_1,a_2)=(1,-2)) = \frac{1}{2}$. Then we get

$$X_0(\alpha) = \{x \in \mathbb{R}^2 : 2x_1 - x_2 \leq 0 \text{ and } 2x_2 - x_1 \leq 0\}, 0 \leq \alpha < \frac{1}{2}.$$

$$X_0(\alpha) = \{x \in \mathbb{R}^2 : 2x_1 - x_2 \leq 0 \text{ or } 2x_2 - x_1 \leq 0\}, \frac{1}{2} \leq \alpha < 1.$$

$$X_1(\beta) = \{x \in \mathbb{R}^2 : 2x_1 - x_2 \leq 2\beta, 2x_2 - x_1 \leq 2\beta \text{ and } x_1 + x_2 \leq 2\beta \}, 0 \leq \beta < \infty.$$

$$X_2(\alpha) = \{x \in \mathbb{R}^n : (2-3\alpha)x_1 + (3\alpha - 1)x_2 \leq 0 \text{ and } (2-3\alpha)x_2 + (3\alpha - 1)x_1 \leq 0 \}, 0 \leq \alpha \leq \frac{1}{2}.$$

$$X_2(\alpha) = \{x \in \mathbb{R}^n : (2-3\alpha)x_1 + (3\alpha - 1)x_2 \leq 0 \text{ or } (2-3\alpha)x_2 + (3\alpha - 1)x_1 \leq 0 \}, \frac{1}{2} \leq \alpha < 1.$$

In contrary to $X_0(\alpha)$, $X_2(\alpha)$ increases strictly and continuously for $\alpha \in (0,1)$. Also $X_2(\beta)$ increases strictly and continuously for $\beta \in (0,\infty)$. $X_1(\beta)$ is convex $\forall \beta$, $X_2(\alpha)$ is convex $\forall \alpha \leq \frac{1}{2}$.

Example 2. n = 1, $P(b=0) = 1$, $P(a_1=-1) = 2/3$, $P(a_1=+1) = 1/3$.

$$X_0(\alpha) = X_2(\alpha) = \{0\} , 0 \leq \alpha < 1/3,$$

$$= (-\infty,0] , 1/3 \leq \alpha < 2/3,$$

$$= \mathbb{R} , 2/3 \leq \alpha \leq 1.$$

$$X_1(\beta) = [-3\beta, 3\beta/2] , \beta \geq 0.$$

Now $X_1(\beta)$ and $X_2(\alpha)$ are convex for all α and β. $X_1(\beta)$ increases strictly and continuously, but $X_2(\alpha)$ does not!

Remark. Most of the results in theorems 2 and 3 are still true, if the underlying constraint (1) is replaced by the *nonlinear* constraint for $x \in \mathbb{R}^n$

(1') $g(x,\omega) \geqq 0$

where ω is a random vector with support Ω. If it is assumed that

$g(.,\omega)$ is concave for all $\omega \in \Omega$,

(8') $E_\omega |g(x,\omega)| < \infty$ for all $x \in \mathbb{R}^n$ and

(17') $E_\omega |g(\hat{x},\omega)| > 0$ for at least one $\hat{x} \in \mathbb{R}^n$

then again $X_1(\beta)$, $\beta \geq 0$, and $X_2(\alpha)$, $0 \leq \alpha \leq \frac{1}{2}$, are closed convex sets, strictly and continuously increasing with the risk aversion parameters β and α, respectively. Of course, here we identify $\eta(x)$ with $g(x,\omega)$. The proof is a repetition of the proof of the corresponding statements in theorems 2 and 3, using the fact that also in the more general case the functions $E[\eta(x)]^-$ and $-E(\eta(x))$ are finite, convex and continuous.

It is important to notice, that the results of theorems 2 and 3 do not depend on the type of the distribution of $(a_1,\ldots,a_n,b)$. This is a basic difference between CC and ICC. For example, the convexity of $X_0(\alpha)$ has only been established for special distributions (e.g. $(a_1,\ldots,a_n)$ deterministic, or $(a_1,\ldots,a_n,b)$ having a log-concave density [9]), whereas it is nonconvex on a large range of α, generally, if the distribution of $(a_1,\ldots,a_n,b)$ is discrete. Also, $X_0(\alpha)$ is only continuously and strictly increasing with α if $(a_1,\ldots,a_n,b)$ has a density with connected support; its increase is stepwise if the distribution is discrete (as in the examples 1 and 2).

It is clear that convexity of the feasible set is an important property for computational reasons. Also from a modeling point of view it may be of value to maintain the property 'a combination of acceptable solutions is acceptable too' in the case of stochastic linear constraints ([7]; [13] p.573); of course, this property is fundamental in the deterministic case. In practice the specification of the numerical values for the risk parameters α, β is often more or less arbitrary. For the sensitivity analysis with respect to them the results on strict and continuous increase of the ICC feasibility sets may be of use. For extreme specifications of α and β the ICCs agree completely with the CCs: $X_0(0) = X_1(0) = X_2(0) = \{x \in \mathbb{R}^n: P(\eta(x) < 0) = 0\}$, the 'extremely safe' set, whereas $X_0(1) = X_1(\infty) = X_2(1) = \mathbb{R}^n$. Defining the 'extremely risky' set R by

$$R = \{x \in \mathbb{R}^n : P(\eta(x) < 0) = 1\},$$

it is easily seen, that $X_0(\alpha) \cap R = X_2(\alpha) \cap R = \emptyset$ for all $\alpha < 1$. However, it is quite possible that $X_1(\beta) \cap R \neq \emptyset$, even for small values of β (e.g. take $x_1 = x_2 = \beta$ in example 1). This might be quite acceptable in practical applications: if $x \in X_1(\beta) \cap R$ one is sure that positive shortage will appear, but also that its mean value is small enough.

The choice $\alpha = \frac{1}{2}$ in $X_0(\alpha)$ reflects more or less a risk-neutral attitude: the median of $\eta(x)$ must be nonnegative. Similarly, $\alpha = \frac{1}{2}$ in $X_2(\alpha)$ corresponds to a risk-neutral attitude: the mean value of $\eta(x)$ must be nonnegative. In fact, the practice of neglecting randomness in the constraints may be interpreted as replacing (1) by $X_2(\frac{1}{2})$. Consequently, in $X_2(\alpha)$ one should specify $\alpha \leq \frac{1}{2}$ just as in usual CC.

It is well-known that in special cases CCs have easy deterministic equivalents. The same is true for ICC. Suppose first that $(a_1,\ldots,a_n)$ is deterministic. Then $E[\eta(x)]^- = g(\Sigma_{j=1}^n a_j x_j)$ with $g(z) = E[z-b]^-$. By partial integration one derives, using the finiteness of $\bar{b} = E(b)$,

$$(25) \qquad g(z) = \int_z^\infty (1-F(t))dt = \bar{b} - z + \int_{-\infty}^z F(t)dt , \quad z \in \mathbb{R},$$

where $F(t) := P(b < t)$. The function g is convex, continuous, with subgradient $\partial g(z) = [-1+P(b<z), -1+P(b \leq z)]$. Its asymptotes for $|z| \to \infty$ are given by $g_0(z) = [z-\bar{b}]^-$; in fact, $g(z) = g_0(z)$ iff $z \leq b_0$ or $z \geq b_1$, where $b_0 :=$ inf $\{t: F(t) > 0\}$ and $b_1 :=$ sup $\{t: F(t) < 1\}$. From these characterization of the function g one easily derives deterministic equivalents for $X_1(\beta)$ and $X_2(\alpha)$; they have the form $\{x: \Sigma_{j=1}^n a_j x_j \geq k\}$ for a certain k, as is well-known for $X_0(\alpha)$.

Theorem 4. (*Only right hand side random.*)

$X_0(\alpha) = \{x \in \mathbb{R}^n : \Sigma_{j=1}^n a_j x_j \geq k_\alpha^0\}$, $0 < \alpha < 1$,

where $k_\alpha^0 := $ inf $\{z: -\alpha \in \partial g(z)\}$ (i.e. $k_\alpha^0 \in F^{-1}(1-\alpha)$).

$X_1(\beta) = \{x \in \mathbb{R}^n : \Sigma_{j=1}^n a_j x_j \geq k_\beta^1\}$, $0 < \beta < \infty$,

where k_β^1 is the unique solution of $g(k) = \beta$.

$X_2(\alpha) = \{x \in \mathbb{R}^n : \Sigma_{j=1}^n a_j x_j \geq k_\alpha^2\}$, $0 < \alpha < 1$,

where k_α^2 is the unique solution of $(1-2\alpha)g(k) = \alpha(k-\bar{b})$.

The functions $\beta \to k_\beta^1$, $0 < \beta < \infty$ and $\alpha \to k_\alpha^2$, $0 < \alpha < 1$, are continuous and strictly decreasing, and

$$\lim_{\alpha \downarrow 0} k_\alpha^0 = \lim_{\beta \downarrow 0} k_\beta^1 = \lim_{\alpha \downarrow 0} k_\alpha^2 = b_1 ,$$

$$\lim_{\alpha \uparrow 1} k_\alpha^0 = \lim_{\alpha \uparrow 1} k_\alpha^2 = b_0 ,$$

$$\lim_{\beta \uparrow \infty} (k_\beta^1 - (\bar{b} - \beta)) = 0 .$$

Moreover, $k_{\frac{1}{2}}^2 = \bar{b}$.

Proof. The formula for $X_2(\alpha)$ is based on (16c). The statements about k_α^0, k_β^1, k_α^2 follow directly from the characterization of g, or from theorems 2 and 3. □

Suppose now that $(a_1, \ldots, a_n, b)$ are normally distributed. In this case $\eta(x)$ has also a normal distribution with mean value $\mu(x) = \sum_{j=1}^{n} \bar{a}_j x_j - \bar{b}$, and variance $\sigma^2(x) = \tilde{x}' \Sigma \tilde{x}$, with $\tilde{x} = (x,1)$, $\Sigma =$ covariance matrix of (a,b). It is well-known (e.g. [5]) that $\sigma(x)$ is a convex function. If $\sigma(x) = 0$ then $E[\eta(x)]^- = [\mu(x)]^-$, and if $\sigma(x) > 0$ $E[\eta(x)]^- = \sigma(x) g_1(\frac{\mu(x)}{\sigma(x)})$, where $g_1(z) :=$ $E[z-u]^-$, u being a standard normal random variable. Obviously, the analysis of g_1 follows from that of g; from (25) we get

(26) $g_1(z) = \int_z^\infty (1 - \Phi_0(t)) dt = -z + \int_{-\infty}^z \Phi_0(t) dt$

Φ_0 and φ_0 being the distribution function and the density function of the standard normal distribution. Since $\varphi_0'(z) = -z \varphi_0(z)$ it follows that $\frac{d}{dz}(z \Phi_0(z) + \varphi_0(z)) = \Phi_0(z)$, so that $g_1(z)$ can be expressed as

(27) $g_1(z) = -z + z \Phi_0(z) + \varphi_0(z)$, $z \in \mathbb{R}$.

Theorem 5. (Normal distributions.)

$X_0(\alpha) = \{x \in \mathbb{R}^n : \mu(x) \geq k_\alpha^0 \cdot \sigma(x)\}$, $0 < \alpha < 1$,

where $k^0 = \Phi_0^{-1}(1-\alpha)$ (i.e. $g_1'(k_\alpha^0) = -\alpha$).

$X_1(\beta) = \{x \in \mathbb{R}^n : \mu(x) \geq k_{\beta/\sigma(x)}^1 \cdot \sigma(x)\}$, $0 < \beta < \infty$,

where k_γ^1 is the unique solution of $g_1(k) = \gamma$.

$X_2(\alpha) = \{x \in \mathbb{R}^n : \mu(x) \geq k_\alpha^2 \cdot \sigma(x)\}$, $0 < \alpha < 1$,

where k_α^2 is the unique solution of $(1-2\alpha) g_1(k) = \alpha k$.

$X_1(\alpha)$ and $X_2(\alpha)$ are convex if $\alpha \leq \frac{1}{2}$; $X_1(\beta)$ is convex for all β. The functions $\alpha \to k_\alpha^0$, $\beta \to k_\beta^1$, $\alpha \to k_\alpha^2$ are continuous and strictly decreasing, with $k_{\frac{1}{2}}^0 = k_{\frac{1}{2}}^2 = 0$, $k_\gamma^1 = 0$ if $\gamma = (2\pi)^{-\frac{1}{2}}$ and

$$\lim_{\alpha \downarrow 0} k_\alpha^0 = \lim_{\gamma \downarrow 0} k_\gamma^1 = \lim_{\alpha \downarrow 0} k_\alpha^2 = +\infty$$

$$\lim_{\alpha \uparrow 1} k_\alpha^0 = \lim_{\alpha \uparrow 1} k_\alpha^2 = -\infty, \quad \lim_{\gamma \to \infty} (k_\gamma^1 + \gamma) = 0.$$

Proof. As the proof of theorem 4. □

Remark. Up to now we considered ICC formulations for an inequality constraint (1). Whereas the CC formulation of an *equality constraint* $\sum_{j=1}^n a_j x_j = b$ is useless, this is not true for ICCs. For example one may define

(28) $X_4(\beta_1, \beta_2) := \{x \in \mathbb{R}^n : E[\eta(x)]^- \leq \beta_1, E[\eta(x)]^+ \leq \beta_2\}$,

(29) $X_5(\beta) \quad := \{x \in \mathbb{R}^n : E|\eta(x)| \leq \beta\}$.

Both sets are convex, but they are empty if the risk aversion parameters are specified too small.

§3. Joint Integrated Chance Constraints. In this section we consider $m \geq 2$ random constraints

(30) $A_i x := \sum_{j=1}^n a_{ij} x_j \geq b_i$, $i = 1, \ldots, m$,

where $(A, b) = (a_{ij}, b_i)$ has a known distribution with finite means. If for each constraint a separate degree of risk aversion is specified, (30) can be reformulated in terms of ICCs as

(31) $x \in \tilde{X}_1(\beta_1, \ldots, \beta_m) := \cap_{i=1}^m X_{1i}(\beta_i)$, $0 \leq \beta_i < \infty$,

(32) $x \in \tilde{X}_2(\alpha_1, \ldots, \alpha_m) := \cap_{i=1}^m X_{2i}(\alpha_i)$, $0 \leq \alpha_i \leq 1$,

where $X_{1i}(\beta_i)$ and $X_{2i}(\alpha_i)$ are defined as in (3) and (6). Of course, $\tilde{X}_1$ and $\tilde{X}_2$ have the same properties as X_1 and X_2.

A well-known alternative for separate CCs is the *joint* CC

(33) $Y_0(\alpha) := \{x \in \mathbb{R}^n : P(Ax \geq b) \geq 1-\alpha\}$, $0 \leq \alpha \leq 1$.

Since $Y_0(\alpha)$ can be formulated as

(34) $Y_0(\alpha) = \{x \in \mathbb{R}^n : E \text{ sgn } \max_i [\eta_i(x)]^- \leqq \alpha\}$

where $\eta_i(x) := \Sigma_{j=1}^n a_{ij} x_j - b_i$, the obvious generalization of $X_1(\beta)$ to joint ICC is

(35) $Y_1(\beta) := \{x \in \mathbb{R}^n : E \max_i [\eta_i(x)]^- \leqq \beta\}, \ 0 \leqq \beta < \infty.$

Unlike $Y_0(\alpha)$, $Y_1(\beta)$ is *convex* for all distributions of (A,b) and for all values of the risk aversion parameter. In fact, the same statement is true for

$$Y_1'(\beta) := \{x \in \mathbb{R}^n : E\|[\eta(x)]^-\| \leqq \beta\}, \ 0 \leqq \beta < \infty,$$

where $\|.\|$ denotes *any* norm in $\mathbb{R}^n$. Joint variants of $X_2(\alpha)$ are less obvious. One might define e.g.

$$Y_2(\alpha) := \{x \in \mathbb{R}^n : E \max_i [\eta_i(x)]^- \leqq \alpha E\max_i |\eta_i(x)|\}, \ 0 \leqq \alpha \leqq 1,$$

$$Y_2'(\alpha) := \{x \in \mathbb{R}^n : (1-\alpha) E \max_i [\eta_i(x)]^- \leqq \alpha E \max_i [\eta_i(x)]^+\}, \ 0 \leqq \alpha \leqq 1,$$

$$Y_2''(\alpha) := \{x \in \mathbb{R}^n : (1-2\alpha) E \max_i [\eta_i(x)]^- \leqq \alpha E \min_i (\eta_i(x))\}, \ 0 \leqq \alpha \leqq 1,$$

each of which   reduces to $X_2(\alpha)$ if $m = 1$. $Y_2(\alpha)$ and $Y_2'(\alpha)$ aré nonconvex generally, but $Y_2''(\alpha)$ is convex for $\alpha \leqq \frac{1}{2}$.

It is also possible to define ICC feasibility sets for which no CC companion exists. For example, for weights $r_i > 0$ with $\Sigma_{i=1}^m r_i = 1$

(36) $Z_1(\beta,r) := \{x \in \mathbb{R}^n : \Sigma_{i=1}^m r_i E[\eta_i(x)]^- \leqq \beta\}, \ \beta \geqq 0,$

convex for all $\beta \geqq 0$, and

(37) $Z_2(\alpha,r) := \{x \in \mathbb{R}^n : \Sigma_{i=1}^m r_i E[\eta_i(x)]^- \leqq \alpha \Sigma_{i=1}^m r_i E|\eta_i(x)|\}, \ 0 \leqq \alpha \leqq 1,$

convex for all $\alpha \leqq \frac{1}{2}$.

§4. <u>Relation between Integrated Chance Constraints and Recourse Models</u>. Just as in ICCs, in simple recourse models 'risk' is interpreted as 'mean shortage'. But risk aversion is specified in terms of penalty costs for shortages rather than by prescribing the maximal acceptable risk. It is clear, that there must be an intimate relation between both model types, expressed by Lagrange multipliers. It will be shown that for each convex ICC model an equivalent (often simple) recourse model exists. Hence, the criticism in [4], that CCP is a deficient modeling technique because of the poor correspondence between CCP and SPR, is is not valid for ICCs.

Suppose that the finite convex objective function $c(x)$ is to be minimized on the nonempty convex set X. Then (31) is part of the problem

$$ICC_1(\beta): \min_{x \in X}\{c(x): E[A_i x-b_i]^- \le \beta_i, \ i = 1,\dots,m\}, \ \beta \ge 0.$$

This convex program is strictly feasible if β is not too small, that is: $\beta_i > \beta_{i0} := E[A_i \hat{x}-b_i]^-$, $\forall i$, for a suitable $\hat{x} \in X$. The Lagrangian problem

$$L_1(\lambda): \min_{x \in X}\{c(x)+\Sigma_{i=1}^m \lambda_i \cdot E[A_i x-b_i]^-\}, \ \lambda \ge 0,$$

is a simple recourse model, and we have

Theorem 6. ($ICC_1(\beta)$ and $L_1(\lambda)$ are equivalent.)
(a) If $\beta_i > \beta_{i0} \forall i$ then there exist optimal Lagrange multipliers λ_i^0 for the constraints in $ICC_1(\beta)$, and $x^* \in X$ solves $ICC_1(\beta)$ iff it is feasible for it and is a solution of $L_1(\lambda^0)$.
(b) If $x^* \in X$ solves $L_1(\lambda)$ for any $\lambda \ge 0$ then it solves $ICC_1(\beta)$ with $\beta_i := E[A_i x^*-b_i]^-$.

Proof. Follows directly from the Kuhn-Tucker theorem; see e.g. [10] Thms. 28.2 and 28.3.   □

As said before, the equivalence in theorem 6 does not mean that both models are equivalent from a practical point of view, since the specification of penalty costs may be more difficult or less difficult than that of the risk aversion parameters β.

Also for separate ICCs based on (32), one proves similarly that

$$ICC_2(\alpha): \min_{x \in X}\{c(x): E[A_i x-b_i]^- \le \alpha_i E|A_i x-b_i|, \ i = 1,\dots,m\}, \ 0 \le \alpha_i \le \tfrac{1}{2},$$

is related (in the sense of theorem 6a) to the simple recourse model

$$L_2(\lambda): \min_{x \in X}\{c(x)+\Sigma_{i=1}^m ((1-\alpha_i)\lambda_i E[A_i x-b_i]^- + (-\alpha_i)\lambda_i E[A_i x-b_i]^+), \ \lambda \ge 0.$$

Similar for (37). For (28), (29) and (36) one easily formulates equivalent (i.e. theorem 6a and 6b holds) simple recourse models.

Also for the joint integrated chance constraints (35) equivalence to a recourse model can be proved. In this case the recourse is fixed and complete, but not simple. Defining

$$ICC_3(\beta): \min_{x \in X}\{c(x): \; E\max_i[A_i x - b_i]^- \leq \beta\}, \; \beta \geq 0,$$

and

$$L_3(\lambda): \min_{x \in X}\{c(x) + Q_\lambda(x)\}, \; \lambda \geq 0,$$

$$Q_\lambda(x) := E \min_y \{qy: \; Wy = b - Ax, \; y \geq 0\}, \text{ where}$$

$$\binom{q}{W} := \binom{\lambda \quad 0}{e \quad -I}, \; e \in \mathbb{R}^m \text{ with } e_i = 1 \; \forall i,$$

one easily verifies that $Q_\lambda(x) = \lambda.E\max_i[A_i x - b_i]^-$, so that theorem 6 holds also for $ICC_3(\beta)$ and $L_3(\lambda)$.

§5. Conclusions. In many cases, ICCs may be an appropriate tool for modeling random linear constraints. ICCs are more appropriate than CCs if the underlying concept 'risk := mean shortage' is more appropriate than 'risk:= probability of positive shortage'. ICCs give rise to convex optimization problems, for all distributions of the random coefficients involved, matrix as well as righthand side. Moreover, the feasibility sets defined by ICCs change continuously and strictly for all types of distributions of the random coefficients. Models with ICCs are more appropriate that SPR models, if one is not able to specify the penalty costs for infeasibilities. On the other hand, there is a natural mathematical equivalence between ICC and SPR models, provided by Lagrange multipliers. Computation of mean shortages is not easy in general; however, in simple situations where deterministic equivalent formulations for CCs exist, the same is true for ICCs, and as far as Monte Carlo simulation is concerned, there does not seem to be much difference in difficulty either. On the other hand, one might expect that progress in computation in simple and fixed recourse models can be used for computation with ICCs.

References.

[1] DEMPSTER, M.A.H., "On Stochastic Programming: I. Static Linear Programming under Risk", J. Math. Anal. Applns, 21(1968)304-343.
[2] GARSTKA, S.J., "The Economic Equivalence of Several Stochastic Programming Models", in Stochastic Programming, Dempster, M.A.H. (ed.), Academic Press, New York, 1980, pp. 83-91.

[3] GARSTKA, S.J. AND WETS, R.J.-B., "On Decision Rules in Stochastic Program-
       ming", *Math. Programming* 7(1974)117-143.
[4] HOGAN, A.J., MORRIS J.G. AND THOMPSON, H.E., "Decision Problems under
       Risk and Chance Constrained Programming: Dilemmas in the Transition,"
       Man. Science 27(1981)698-716.
[5] KALL, P., "Stochastic Linear Programming", Springer Verlag, Berlin,
       Heidelberg, New York, 1976.
[6] KALL, P. AND STOYAN, D.J., "Solving Stochastic Programming Problems with
       Recourse Including Error Bounds", *Math. Operationsforsch. Stat.*
       13(1982)431-447.
[7] KLEIN HANEVELD, W.K., "Alternatives for Chance Constraints: Integrated Chance
       Constraints", Report 97(OR-8301), Econometric Inst., Univ. of Groningen, 1983.
[8] PREKOPA, A., "Contributions to the Theory of Stochastic Programming,", *Math.*
       Programming 4(1973)202-221.
[9] PREKOPA, A., "Programming under Probabilistic Constraints with a Random
       Technology Matrix, *Math. Operationsforsch. Stat.* 5(1974)109-116.
[10] ROCKAFELLAR, R.T., "Convex Analysis", Princeton University Press, Princeton
       N.J., 1970.
[11] SYMONDS, G.H., "Chance-Constrained Equivalents of Some Stochastic Programming
       Problems", *Operations Research* 16(1968) 1152-1159.
[12] WALKUP, D.W. AND WETS, R.J.-B., "Stochastic Programs with Recourse: Special
       Forms", in *Proceedings of the Princeton Symposium on Mathematical*
       Programming, Kuhn, H. (ed.), Princeton University Press, Princeton, N.J.
       1970, pp. 139-161.
[13] WETS, R., Stochastic Programming: Solution Techniques and Approximation
       Schemes", in *Mathematical Programming, The State of the Art*, Bachum, A.,
       Grötschel, B. and Korte, B., (eds.), Springer Verlag, Berlin, Heidelberg,
       New York, Tokyo (1983) pp. 566-603.
[14] WILLIAMS, A.C., "On Stochastic Linear Programming", *SIAM J. Appl. Math.*
       13(1965)927-940.

ALGORITHMS BASED UPON GENERALIZED LINEAR PROGRAMMING FOR STOCHASTIC PROGRAMS WITH RECOURSE

J.L. Nazareth

IIASA

A - 2361 Laxenburg, Austria

1. INTRODUCTION

We are concerned here with two-stage stochastic linear programs (SLP) with recourse, of the form

minimize $cx + Q(x)$

subject to

$$Ax = b$$

$$x \geq 0 \tag{1.1a}$$

where

$$Q(x) = E\{Q(x,h(w))\} \tag{1.1b}$$

and

$$Q(x,h(w)) = \inf_{y \geq 0}\{qy \mid Wy = h(w) - Tx\} \tag{1.1c}$$

In the above, only the right-hand-side $h(w)$, is a random vector defined on a probability space whose events are denoted by w. E denotes expectation. T denotes the fixed $m_2 \times n_1$ technology matrix and W the fixed $m_2 \times n_2$ recourse matrix. A is an $m_1 \times n_1$ matrix defining the constraints, and c,b,q,x,y are vectors of appropriate dimension. We shall be concerned with problems of the form (1.1a-c) with *complete recourse* i.e. with constraints which satisfy

$$\text{pos } W \equiv \{t \mid t = Wy, \; y \geq 0\} = \mathbb{R}^{m_2}. \tag{1.1d}$$

Since T is fixed, we can define the (non-stochastic) *tender* $\chi = Tx$ and write (1.1a-c) in the equivalent form:

minimize $cx + \Psi(\chi)$

subject to

   $Ax = b$

   $Tx - \chi = 0$ (1.2a)

   $x \geq 0$

where

   $\Psi(\chi) = E\{\psi(\chi, h(w))\}$ (1.2b)

and

   $\psi(\chi, h(w)) = \inf\limits_{y \geq 0}\{qy \mid Wy = h(w) - \chi\}$ (1.2c)

   We show first that an equivalent form to (1.2a) is

minimize $cx + qy + \Psi(\chi)$

subject to

   $Ax = b$

   $Tx + Wy - \chi = 0$

   $x, y \geq 0.$ (1.3)

   The family of algorithms that we are concerned with here were introduced in Nazareth and Wets, 1983, and are based upon the generalized linear programming (GLP) method of Wolfe (see Dantzig, 1963, Shapiro, 1979). They successively inner linearize $\Psi(\chi)$ in (1.3) and solve a sequence of master linear programming problems of the form

minimize $cx + qy + \sum\limits_{k=1}^{K} \lambda_k \Psi(\chi^k)$

subject to

$$Ax = b \tag{1.4}$$

$$Tx + Wy - \sum_{k=1}^{K} \lambda_k \chi^k = 0$$

$$\sum_{k=1}^{K} \lambda_k = 1$$

$$x, y, \lambda_k \geq 0$$

The *tenders* $\chi^1, \ldots, \chi^K$ are assumed to have been previously generated and at the current cycle of the algorithm a new tender χ^{K+1} is
introduced by solving the (Lagrangian) subproblem

$$\text{minimize } \Psi(\chi) + \pi^K \chi \tag{1.5}$$
$$\chi \in X$$

where π^K are the dual multipliers associated with the constraints

$Tx - \sum_{k=1}^{K} \lambda_k \chi^k = 0$ in the optimal solution of (1.4). χ^{K+1}, the optimal

solution$^{(*)}$ of (1.5), is an improving tender provided that
$\Psi(\chi^{K+1}) + \pi^K \chi - \theta^K < 0$, where θ^K is the optimal dual multiplier associated with the constraint $\sum_{k=1}^{K} \lambda_k = 1$. When χ^{K+1} is introduced into
the master problem (1.4), such a tender will lead to a reduction in
the objective value (barring degeneracy, of course). Since the projection of the set of vectors (x,y,χ) satisfying $Ax = b$, $Tx + Wy - \chi = 0$,
$x,y \geq 0$ onto the space of the χ vectors is $\mathbb{R}^{m_2}$ by (1.1d), χ can be
assumed unrestricted in (1.5). However, it is often convenient to con-
fine χ to some compact set X defined by simple bounds, for reasons of
computational efficiency and to facilitate convergence arguments. Extensions to include lines of recession in (1.4) and relax the restriction (1.1d) will not be considered in this paper.

When the recourse is *simple* i.e., when $W = [I,-I]$, an approach
based upon generalized linear programming has been suggested more than
one in the literature, see, for example, Williams, 1966, Parikh, 1968.
However, apart from special applications, see Ziemba, 1972, it has
not been pursued in any real computational way. For problems with
general recourse it has apparently not beén tried at all. Moreover,
it is important to recognize that the GLP approach should be combined
with a suitable *problem transformation*, for example, the one involved
in going from (1.1a-c) to (1.2a-c), in order to keep the degree of

---

(*) In practice (1.4) does not have to be pushed to optimality at each
    iteration, but this is a question of strategy, which we discuss
    later.

nonlinearity low. This was not fully appreciated, at least from an algorithmic point of view.

We turn now to the organization of our paper. In Section 2, we consider the alternative formulation of the equivalent deterministic form (1.2a), given by (1.3) and an interpretation of the solution of the above algorithm (1.4) and (1.5). In particular, we wish to see how *tenders* and *certainty equivalents* stand in relation to one another. Next we consider problems with simple recourse. We discuss algorithms for two cases: a) When the distribution is discrete and probabilities are known explicitly . Then $\Psi(\chi)$ is much more tractable. b) When the probability distribution is other than discrete or when it is only known implicitly through some simulation model involving the random elements w.

Case b) above is especially useful because it enables us to make the transition to general recourse, which is the topic of Section 4. Here $\Psi(\chi)$ is usually difficult to compute, since it involves minimization calculations and an integration. Our aim in this section is to discuss some possible solution strategies based upon generalized programming. Finally, Section 5 contains some concluding remarks.

Henceforth in this paper when, for example, the text includes equations (1.1a),(1.1b),(1.1c),(1.1d) and we refer to (1.1), we are making reference to all four equations.

2. EQUIVALENT FORMS AND AN INTERPRETATION OF THE SOLUTION

The notion of *certainty equivalent* of a SLP with recourse is well known, see Wets, 1974. Here we wish to investigate the tie between *tenders* and *certainty equivalents*, and with this in mind we first consider an alternative form for (1.2). This also turns out to be useful when formulating algorithms, as we shall see later in Section 3.

Suppose, just for the purpose of discussion, that h(w) is replaced by some deterministic quantity, for example its expected value $\bar{h}$. Then to solve this simplified optimization problem, we need only solve a single stage program of the form:

minimize cx + qy

subject to

$$Ax = b$$

$$Tx + Wy - \bar{h} = 0 \tag{2.1}$$

$$x,y \geq 0$$

Indeed, to test the feasibility and boundedness of the original SLP (1.1) we should solve problems of this form for suitably chosen $\bar{h}$, as shown by Wets, 1972.

Upon comparing (2.1) and (1.2), it is tempting to include the recourse matrix W *explicitly* in the first stage i.e., to consider the implications of having the recourse activities available to the first stage. This would often be the case in practice as pointed out by Williams, 1966. We would then have an equivalent deterministic problem of the form:

minimize $cx + qy + \Psi(\chi)$

subject to

$$Ax = b$$

$$Tx + Wy - \chi = 0 \tag{2.2}$$

$$x, y \geq 0$$

with $\Psi(\chi)$ defined by (1.2b-c). We now want to show that (1.2) and (2.2) are equivalent forms.

Let us demonstrate this for the case when h(w) is discretely distributed. Suppose, therefore, that the distribution of h(w) is defined by vectors

$$h^1, h^2, \ldots, h^t \tag{2.3a}$$

with associated probabilities

$$f_1, f_2, \ldots, f_t, \text{ where } \sum_{k=1}^{t} f_k = 1, f_k \geq 0 \tag{2.3b}$$

Then (1.2) can be expressed as follows:

minimize $cx + f_1 qy^1 + f_2 qy^2 + \ldots + f_t qy^t$

subject to

$$
\begin{array}{ll}
Ax & = b \\
Tx + Wy^1 & = h^1 \\
Tx + \quad Wy^2 & = h^2 \\
\quad \vdots \quad \vdots \quad \vdots \\
Tx + \qquad\qquad Wy^t = h^t
\end{array}
\qquad (2.4)
$$

$x, y^j \geq 0$

and (2.2) can be expressed as

minimize $cx + qy + f_1 qy^1 + f_2 qy^2 + \ldots + f_t qy^t$

subject to

$$
\begin{array}{ll}
Ax & = b \\
Tx + Wy + Wy^1 & = h^1 \\
Tx + Wy + \quad Wy^2 & = h^2 \\
\quad \vdots \qquad \vdots \qquad \vdots \\
Tx + Wy + \qquad\quad Wy^t = h^t
\end{array}
\qquad (2.5)
$$

$x, y, y^j \geq 0$

Any feasible solution of (2.4) gives a feasible solution of (2.5), simply by setting $y = 0$. Conversely, by writing $qy = \sum_{k=1}^{t} f_k (qy)$, and regrouping terms in (2.5) we obtain:

minimize $cx + f_1 q(y+y^1) + f_2 q(y+y^2) + \ldots + f_t q(y+y^t)$

subject to

$$Ax \qquad\qquad\qquad\qquad = b$$

$$Tx + W(y+y^1) \qquad\qquad = h^1$$

$$Tx + \qquad W(y+y^2) \qquad\qquad = h^2$$

$$\begin{array}{ccc} . & . & . \\ . & . & . \\ . & . & . \end{array} \qquad\qquad (2.6)$$

$$Tx + \qquad\qquad W(y+y^t) = h^t$$

$$x, y, y^j \geq 0$$

and thus any feasible solution of (2.5) gives a feasible solution to (2.4), with the same objective value. The two problems must therefore be equivalent. We are led to the following theorem, a generalization of a result for *simple* recourse given in Parikh, 1968.

   THEOREM 2.1. The SLP problem with recourse given by (1.2) and (2.2) are equivalent, in the following sense:

$$(\bar{x},\bar{\chi}) \quad \text{solves} \quad (1.2) \Rightarrow (\bar{x},0,\bar{\chi}) \quad \text{solves} \quad (2.2)$$

$$(\bar{x},\bar{y},\bar{\chi}) \text{ solves } (2.2) \Rightarrow (\bar{x},\bar{\chi}-W\bar{y}) \text{ solves } (1.2)$$

We assume that (1.2) is solvable (bounded and solution attained); it will imply that (2.2) is solvable, and vice-versa.

   PROOF. [*]

1.   Suppose $\bar{x} \in R_+^{n_1}$, $\bar{y} \in R_+^{n_2}$, $\bar{\chi} \in R_+^{m_2}$ satisfy

$$T\bar{x} + W\bar{y} = \bar{\chi}$$

Let

$$\chi^0 = \bar{\chi} - W\bar{y} = T\bar{x}$$

Then for all $h(\cdot)$

$$\psi(\chi^0, h(\cdot)) \leq \psi(\bar{\chi}, h(\cdot)) + q\bar{y}$$

---

[*] The formal proof of this proposition *for an arbitrary distribution*, which now follows, is due to Roger Wets.

PROOF OF 1. We have to show that

$$\inf_{y \geq 0} (qy \mid Wy = h(\cdot) - \chi^0) \leq q\bar{y} + \inf_{u \geq 0} (qu \mid Wu = h(\cdot) - \bar{\chi})$$

$$= q\bar{y} + \inf_{u \geq 0} (qu \mid Wu = h(\cdot) - \chi^0 - W\bar{y})$$

$$= \inf_{u \geq 0} (q(u + \bar{y}) \mid W(u + \bar{y}) = h(\cdot) - \chi^0)$$

$$= \inf_{y \geq \bar{y}} (qy \mid Wy = h(\cdot) - \chi^0) .$$

But that is now evident since $\bar{y} \in R^{n_2}$ and thus the condition $y \geq \bar{y}$ is more constraining than $y \geq 0$ (except if $\bar{y} = 0$). ■

2. Suppose $\bar{x}, \bar{y}, \bar{\chi}, \chi^0$ are as in 1. Then

$$\Psi(\chi^0 = \bar{\chi} - W\bar{y}) \leq \Psi(\bar{\chi}) + q\bar{y}$$

PROOF OF 2. Use 1. + the fact: taking expectations is order preserving. ■

3. Suppose $\bar{x}, \bar{y}, \bar{\chi}$ is any feasible solution of (2.2). Then

$$c\bar{x} + q\bar{y} + \Psi(\bar{\chi}) \geq c\bar{x} + q \cdot 0 + \Psi(\chi^0)$$

where

$$\chi^0 = \bar{\chi} - W\bar{y} = T\bar{x}$$

PROOF OF 3. Follows from 2.; add $c\bar{x}$ on each side. ■

From 3. it follows that in order to find the infimum in (2.2), it suffices to restrict oneself to feasible solutions of (2.2) that have $y = 0$. But then (2.2) is exactly (1.2). Thus if $(\bar{x}, \bar{\chi})$ solves (1.2), the triple $(\bar{x}, 0, \bar{\chi})$ solves (2.2). If $(\bar{x}, \bar{y}, \bar{\chi})$ solves (2.2) and $\bar{z} = c\bar{x} + q\bar{y} + \Psi(\bar{\chi})$ then 3. implies that

$$\bar{z} = c\bar{x} + q \cdot 0 + \Psi(\bar{\chi} + W\bar{y})$$

since the triple $(\bar{x}, 0, \bar{\chi} - W\bar{y})$ is also a feasible solution of (2.2). And the pair $(\bar{x}, \bar{\chi} - W\bar{y})$ solves (1.2) since $(\bar{x}, \bar{\chi} - W\bar{y})$ solves (2.2) when $y(=0)$ is deleted from the problem. This completes the proof of the theorem. ■

In the light of the above proposition, we can deal henceforth with (2.2). Suppose we now apply the GLP algorithm outlined in Section 1 to (2.2). This will give Master LP problems of the form:

$$\text{minimize } cx + qy + \sum_{k=1}^{K} \lambda_k \Psi(\chi^k)$$

subject to

$$Ax = b$$

$$Tx + Wy - \sum_{k=1}^{K} \lambda_k \chi^k = 0 \qquad (2.7)$$

$$\sum_{k=1}^{K} \lambda_k = 1$$

$$x, y, \lambda_k \geq 0$$

Let the optimal solution of (2.7) be x^*, y^*, λ^*, and note that no more than $(m_2 + 1)$ components of λ^* are non-zero. Without loss of generality we can assume that these are the first (m_2+1) components $\lambda_1^*, \ldots, \lambda_{m_2+1}^*$, and we define

$$\chi^* = \sum_{k=1}^{m_2+1} \lambda_k^* \chi^k \qquad (2.8)$$

χ^* is the *certainty equivalent*, since x^* and y^* are optimal for the LP problem

$$\text{minimize } cx + qy$$

subject to

$$Ax = b$$

$$Tx + Wy - \chi^* = 0 \qquad (2.9)$$

$$x, y \geq 0$$

Indeed we can go further. Suppose that we approximate the distribution of $h(w)$ by the following discrete distribution, whose values are

$$x^1, x^2, \ldots, x^{m_2+1} \tag{2.10a}$$

with associated probabilities

$$\lambda_1^*, \lambda_2^*, \ldots, \lambda_{m_2+1}^* \tag{2.10b}$$

where the optimal solution λ^* to (2.7) can be interpeted as defining a probability distribution since

$$\sum_{k=1}^{K} \lambda_k^* = 1, \quad \lambda_k^* \geq 0.$$

For the distribution (2.10), an equivalent form for (1.2) is

$$\text{minimize } cx + \lambda_1^* qy^1 + \lambda_2^* qy^2 + \ldots + \lambda_{m_2+1}^* qy^{m_2+1}$$

subject to

$$
\begin{aligned}
Ax & = b \\
Tx + Wy^1 & = \chi^1 \\
Tx + \quad Wy^2 & = \chi^2 \\
& \quad \vdots \\
Tx + \qquad Wy^{m_2+1} & = \chi^{\cdot m_2+1} \\
x, y^j & \geq 0
\end{aligned}
\tag{2.11}
$$

For any $x \geq 0$ satisfying $Ax = b$, in particular for x^*, we know that (2.11) has a feasible solution for problems with relatively complete recourse. Let $y^{*1}, \ldots, y^{*m_2+1}$ be the corresponding components of the optimal solution of (2.11). Then using Jensen's Inequality, namely $EF(x,\xi) \leq F(x,E\xi)$ we can deduce from the optimal solutions to (2.11) and (2.9) that

$$\sum_{k=1}^{m_2+1} \lambda^* qy^{*k} \leq qy^* \tag{2.12}$$

Now in (2.11), multiply the row involving χ^i by λ_i^* and sum. This leads to

$$Tx + W\left(\sum_{k=1}^{m_2+1} \lambda_k^* y^k\right) - \chi^* = 0 \qquad (2.13)$$

When

$$y = \sum_{k=1}^{m_2+1} \lambda_k^* y^k \ ,$$

we have (x,y) feasible for (2.9), and thus any feasible solution of (2.11) leads to a feasible solution of (2.9). This fact combined with (2.12) implies that (2.9) and (2.11) are equivalent, and we have proved the following theorem which gives an interpretation of the optimal solution of (2.7):

THEOREM 2.2. Suppose that the nonzero components in the optimal solution of (2.7) are given by $\lambda_1^*,\ldots,\lambda_{m_2+1}^*$ with associated tenders $\chi^1,\ldots,\chi^{m_2+1}$, where, without loss of generality, we have assumed these to be the first (m_2+1) components. Then the problem (1.2) is equivalent to the associated discretized problem, obtained by replacing the distribution of $h(w)$ by the distribution (2.10).

3. ALGORITHMS FOR SLP PROBLEMS WITH SIMPLE RECOURSE

3.1. *Discrete Distributions*

For simple recourse, the recourse problem (1.2b) takes the form

$$\psi(\chi,h(w)) = \inf_{y^+ \geq 0, y^- \geq 0} \{q^+ y^+ + q^- y^- \mid [I,-I]\begin{pmatrix} y^+ \\ y^- \end{pmatrix} = h(w)-\chi\} \qquad (3.1)$$

Let $q = q^+ + q^- > 0$. Assume also that $h(w)$ has a discrete distribution, say with the possible values

$$h_{i1}, h_{i2}, \ldots, h_{in_i} \quad \text{where } h_{il} < h_{i,l+1} \qquad (3.2a)$$

with associated probabilities

$$f_{i1}, f_{i2}, \ldots, f_{in_i} \qquad (3.2b)$$

and let

$$\bar{h}_i = E\{h_i(\cdot)\}$$

Then $\Psi(\chi)$ is given by

$$\Psi(\chi) = \sum_{i=1}^{m_2} \Psi_i(\chi_i) \qquad\qquad (3.3a)$$

where

$$\Psi_i(\chi_i) = \max_{l=0,\ldots,n_i} (s_{il}\chi_i + e_{il}) \qquad\qquad (3.3b)$$

and with the convention $\sum_{t=1}^{0} = 0$

$$s_{il} = \left(\sum_{t=1}^{l} f_{it}\right) q_i - q_i^+, \quad 0 \leq l \leq n_i \qquad\qquad (3.4a)$$

$$e_{il} = q_i^+ \bar{h}_i - q_i\left(\sum_{t=1}^{l} p_{it} f_{it}\right), \quad 0 \leq l \leq n_i \qquad\qquad (3.4b)$$

For a proof see Wets, 1983b. Note also that s_{il} form an increasing sequence with

$$-q_i^+ \leq s_{il} \leq q_i^-, \quad 0 \leq l \leq n_i \qquad\qquad (3.5)$$

and e_{il} form a non-increasing sequence.

3.1.1. *Algorithmic Details*. Let us now look at the main ingredients of an algorithm based upon generalized LP for solving the above problem[(*)].

1. *Computing the Objective Functions*: $\Psi(\chi)$ is easily computed from (3.3) and (3.4). The objective function $cx + \Psi(\chi)$ and it is useful to explicitly introduce a scale factor $\rho > 0$, and define the objective to be $cx + \rho\Psi(\chi)$. This is simply a device for parameterizing the objective function of the recourse problem.

2. *Initialization*: Motivated by the results of Section 2, in particular Theorem 2.1, we initially solve the problem

---

[(*)] The algorithm of this section 3.1.1 is quite similar to the one given in unpublished notes by Parikh, 1968.

minimize $cx + \rho q^+ y^+ + \rho q^- y^- + \rho \lambda_1 \Psi(\chi^1)$

subject to

$$Ax = b$$

$$Tx + Iy^+ - Iy^- - \lambda_1 \chi^1 = 0$$

$$\lambda_1 = 1 \tag{3.6}$$

$$x, y^+, y^- \geq 0$$

where

$$\chi^1 \equiv \bar{h} = E\{h(\cdot)\}$$

This is, of course, equivalent to (2.1), since $\lambda_1 \equiv 1$ and $\rho \lambda_1 \Psi(\bar{h})$ is just a constant term, but we prefer (3.6) because it is of the same form as the master program below. From Wets, 1972, we see that success-fully solving (3.6) immediately implies feasibility and boundedness of the original problem.

3. *Solving the Master Program*: This has the form

minimize $cx + \rho q^+ y^+ + \rho q^- y^- + \sum_{k=1}^{K} \lambda_k \rho \Psi(\chi^k)$

subject to

$$\sigma^K: \quad Ax = b$$

$$\pi^K: \quad Tx + Iy^+ - Iy^- - \sum_{k=1}^{K} \lambda_k \chi^k = 0 \tag{3.7}$$

$$\theta^K: \qquad \sum_{k=1}^{K} \lambda_k = 1$$

$$x, y^+, y^-, \lambda_k \geq 0$$

Further initial tenders, other than $\chi^1 = \bar{h}$ could be introduced here. Let $\sigma^K, \pi^K, \theta^K$ denote the optimal multipliers of (3.7). Then the components of π^K satisfy

$$-q_i^+ \leq -\pi_i^K \leq q_i^- \tag{3.8}$$

4. *Solution of the (Lagrangian) Subproblem:* **This is given by**

$$\text{minimize } \Psi(\chi) + \pi^k\chi \qquad \qquad (3.9)$$
$$\chi \in X$$

Let us take $X = R^{m_2}$. Since $\Psi(\chi)$ is separable, we must solve the following for $i = 1,2,\ldots,m_2$

$$\text{minimize } \Psi_i(\chi_i) + \pi_i^K\chi_i \qquad \qquad (3.10)$$
$$\chi_i \in R^1$$

and since $\Psi_i(\chi_i)$ is given by (3.3b), we are dealing in (3.10) with the unconstrained minimization of a piecewise-linear function, and this is easily done.

The optimal solution χ_i^{K+1} satisfies

$$-\pi_i^K \in \partial\Psi_i(\chi_i^{K+1}) \qquad \qquad (3.11)$$

Now from (3.4a) we know that

$$-q_i^+ \leq \partial\Psi_i(\chi_i) \leq q_i^- \qquad \qquad (3.12)$$

for any χ_i in the support of the distribution of $h_i(\cdot)$. It follows from (3.8),(3.11) and (3.12) that χ_i^{K+1} can be found such that

$$h_{i1} \leq \chi_i^{K+1} \leq h_{ik_i} \qquad \qquad (3.13)$$

where h_{i1} are defined by (3.2a).

5. *Adding and Deleting Tenders:* A tender χ^{K+1} is improving for (3.9) provided that

$$\Psi(\chi^{K+1}) + \pi^K\chi^{K+1} - \theta^K < 0. \qquad \qquad (3.14)$$

If no such tender can be found, then the current solution is optimal. Note, in particular, that the subproblem does *not* have to be pushed to optimality. Furthermore, several improving tenders, each satisfying (3.14), could be deduced from one call to the subproblem.

We have not investigated in any detail the question of dropping columns corresponding to tenders from (3.7) when they become out-of-date. In implementations of the related Dantzig-Wolfe decomposition algorithm, see for example Ho, 1974, it is common to drop columns from (3.7), when they have not played a role in the optimal solution for some time and the same strategy could obviously be implemented here. The question is discussed further in Nazareth and Wets, 1983. Much of the theory on dropping cutting planes is also applicable, see, for example, Eaves and Zangwill, 1971.

3.1.2. *Experimental Implementation and Test Example*: We have implemented the above algorithm in an experimental code. Matrices are stored as 2-dimensional arrays and sparsity is not taken into account, so that it can only handle relatively small problems. The master program is solved using the Harwell LP code LA01BD and the subproblems (3.10) are solved by simply finding where $s_{il} + \pi_i^K$ changes sign from negative to positive. A single optimal tender is introduced at each iteration, and all tenders are retained in (3.7). The code was written in Fortran for the Vax 11/780 and validated using the test problems and solutions of Kallberg and Kusy, 1976 and Cleef, 1981.

For an illustrative example, consider the following product-mix problem due to Jim Ho. (Though only a small and highly simplified SLP problem, its full scale version comes from a real life application). The problem involves two products and three ingredients. The variables x_i, y_i, z_i are the amounts of ingredients 1 and 2. The demand for each product is a random variable with known probability distribution. The problem can be summarized as follows:

minimize $x_1 + 2y_1 + 3z_1 + x_2 + 2y_2 + 3z_2 + \Psi(\chi)$

subject to

$$
\begin{array}{l}
\text{A matrix} \left\{
\begin{array}{lll}
\text{Fat/Protein in Product 1:} & .3x_1 + .4y_1 + .2z_1 & \geq 3.3 \\[2mm]
\text{Fat/Protein in Product 2:} & .5y_2 + .6z_2 & \geq 4.0 \\[4mm]
\text{Amt. of Ingredient 1:} & x_1 \quad\quad + x_2 & \leq 15. \\[2mm]
\text{Amt. of Ingredient 2:} & y_1 \quad\quad + y_2 & \leq 12.
\end{array}
\right.
\end{array}
$$

$$
\begin{array}{l}
\text{T matrix} \left\{
\begin{array}{ll}
\text{Amt. of Product 1:} & x_1 + y_1 + z_1 \quad\quad - X_1 = 0 \\[2mm]
\text{Amt. of Product 2:} & x_2 + y_2 + z_2 - X_2 = 0
\end{array}
\right.
\end{array}
$$

$$x_i, y_i, z_i \geq 0 \tag{3.15}$$

The penalties for under and over production are 2.0 and 1.0 units respectively and the probability distribution on demand h(w) is as follows:

product 1	levels	8	10	12
	probs	.25	.5	.25
product 2	levels	15	18	20
	probs	.2	.4	.4

The recourse function $\Psi(\chi)$ is defined by (3.1) where $q^+ = (2.0, 2.0)$ and $q^- = (1.0, 1.0)$.

The following table summarizes the progress of the algorithm

Iteration	First period cost cx	Total cost $cx + \Psi(\chi)$
1	39.	46.06
2	39.	44.75
3	37.	43.575
4	35.9	43.4727
5	35.5	43.4625 optimal

Initial Solution: $x_1 = 6.$, $y_1 = 4.$, $z_1 = 0.1$, $x_2 = 9.$, $y_2 = 8.$, $z_2 = 0.$

Initial Tender: $\begin{pmatrix} 10 \\ 18.2 \end{pmatrix}$

Final Solution: $x_1 = 8.$, $y_1 = 2.25$, $z_1 = 0.$, $x_2 = 7.$, $y_2 = 8.$, $z_2 = 0.$

Final Tender: $0.875 \begin{pmatrix} 10 \\ 15 \end{pmatrix} + 0.125 \begin{pmatrix} 12 \\ 15 \end{pmatrix} = \begin{pmatrix} 10.25 \\ 15 \end{pmatrix}$

An implementation of the algorithm of Section 3.1.1 which is designed to solve reasonably large and sparse SLP problems with simple recourse is given in Nazareth and Wets, 1984. Such problems might typically arise when a given linear program is extended into the domain of SLP with simple recourse by allowing some of its right-hand-side elements to be random variables with known probability distribution; if the SLP arose in this way, the row of the original LP matrix corresponding to stochastic rhs elements would then define the T matrix. These considerations have influenced our design of standardized input formats for SLP problems with recourse, in which a "core" file defining elements of A,T,c,b, bounds and ranges on variables is specified in *standard MPS format*, and a "stochastics" file identifying which rows correspond to the T matrix, and defining distributions and recourse costs is specified in an *MPS-like format*. The implementation is based on the MINOS code of Murtagh and Saunders, 1978.

3.2. *When distribution of* h(w) *is other than discrete, or only known implicitly*

In Section 3.1, the discrete distribution of h(w) was known explicitly and this in turn led to the explicit form $\Psi(\chi)$ given by (3.3) and (3.4). When the distribution of h(w) is not discrete, then $\Psi(\chi)$ is not polyhedral and may be difficult to obtain explicitly. (In some cases it will still however, be possible to obtain $\Psi(\chi)$ quite accurately using numerical integration, in particular one dimensional integration routines when $\Psi(\chi)$ is separable). Even when h(w) has a discrete distribution, this may only be known implicitly, for example, through a simulation model involving the (explicitly) known distributions of the random variables w. When interrogated, this model would produce different observations of h(w) distributed according to its joint probability distribution, but the distribution itself is not explicitly available.

In this section we wish to consider modifications to the algorithm of Section 3.1.1 when the distribution function of h(w) is available in a form that provides samples and when estimates of $\Psi(\chi)$ are obtained from a finite set of such samples. The main modifications involve items 1 and 4, with items 2,3 and 5 remaining unchanged, and they are as follows:

1' *Computing* $\Psi(\chi)$: Suppose the distribution is sampled S times, giving observations $h^1, h^2, \ldots, h^S$. Then a crude estimate of $\Psi(\chi)$ is

$$\psi^E(\chi) = \frac{1}{S} \sum_{k=1}^{S} \psi^E(\chi, h^k) \qquad (3.16a)$$

where

$$\psi^E(\chi, h^k) = \sum_{i:(h_i^k - \chi_i) \geq 0} q_i^+ (h_i^k - \chi_i) - \sum_{i:(h_i^k - \chi_i) < 0} q_i^- (h_i^k - \chi_i) \qquad (3.16b)$$

Estimates of the subgradient $\pi(\chi)$ can also be obtained by

$$\pi_i^E(\chi, h^k) = \begin{cases} -q_i^+ & \text{if } (h_i^k - \chi_i) \geq 0 \\ +q_i^- & \text{if } (h_i^k - \chi_i) < 0 \end{cases} \qquad (3.17a)$$

$$\pi_i^E(\chi) = \frac{1}{S} \sum_{k=1}^{S} \pi_i^E(\chi, h^k) \qquad (3.17b)$$

4' *Solving the (Lagrangian) subproblem*: When minimizing (3.9) with $\Psi(\chi)$ being obtained by (3.16) above, we are dealing with a non-smooth unconstrained function with a fixed level of noise (for fixed sample size). In principle we would need to use methods suggested, for example, by Polyak, 1978 and others. In practice, however, it is possible to employ heuristic methods based upon techniques for smooth problems with good results, see Lemarechal, 1982.

3.2.2. *Results of some experimentation.* We modified the experimental code of section 3.2.2 along the above lines. Using a random number generator which produced pseudo/random numbers r, $0 \leq r \leq 1$, we simulate sampling from the discrete distribution (3.2), by generating a sample, say h^k as follows:

$$h_i^k = h_{it} \quad \text{if } \sum_{l=1}^{t+1} f_{il} > r \geq \sum_{l=1}^{t} f_{il}$$

$\Psi(\chi)$ was obtained by (3.16) with a fixed sample size S. Following Lemarechal, 1982, to solve the subproblem (3.9) we employed the VA13AD Harwell code based on the BFGS update, with *subgradient* estimates (3.17) used in place of the gradient.

Results are summarized in the following table: With sample size 300 for estimates of $\Psi(\chi)$ introduced into the master, and sample size 100 for estimates of $\Psi(\chi)$ and its subgradient used in the unconstrained minimization step, the progress of the algorithm during 8 iterations was as follows:

Iteration	First period cost $c\chi$	Total (estimated) cost $c\chi + \Psi(\chi)$
1	39.	44.17
2	38.14	44.86
3	39.	44.46
4	35.27	43.84
5	37.14	43.53
6	36.12	43.33
7	35.76	42.93
8	36.08	42.928

optimal

Initial Solution: $x_1 = 6.$, $y_1 = 4.$, $z_1 = 0.1$, $x_2 = 9.$, $y_2 = 8.$, $z_2 = 0.$

Initial Tender: $\begin{pmatrix} 10 \\ 18.2 \end{pmatrix}$

Final Solution: $x_1 = 7.62$, $y_1 = 2.54$, $z_1 = 0.$, $x_2 = 7.38$, $y_2 = 8.$, $z_2 = 0.$

Final Tender: $0.927 \begin{pmatrix} 10.02 \\ 15.25 \end{pmatrix} + 0.073 \begin{pmatrix} 11.91 \\ 17.04 \end{pmatrix} = \begin{pmatrix} 10.14 \\ 15.38 \end{pmatrix}$

There are obviously many different strategies that could be used here e.g. progressively increase sample size, and refinement of the estimation of $\Psi^E(\chi)$.

4. GENERAL RECOURSE

In (1.2c), $\Psi(\chi, h(w))$ is now given by the solution of an LP problem defined by W. Since the computation of $\Psi(\chi)$ by (1.2b) involves a multidimensional integration over $\psi(\chi, h(w))$ it is, in general, a function that is difficult to compute.

As in Section 3, we distinguish two cases: a) when $\Psi(\chi)$ and possibly a subgradient of $\Psi(\chi)$ can be computed accurately, in particular, when the distribution of $h(w)$ is defined by a set of scenarios, each having a known probability; b) when $\Psi(\chi)$ and elements of $\partial\Psi(\chi)$ must be approximated in some way. Case b) is much more common, but it pays to dwell on case a), because it gives a lot of insight into methods of solution.

Our aim in this section is to give an overview of some approaches to solving (1.2) based upon generalized linear programming, and not to give specific algorithms.

4.1. Scenarios with known probabilities

Suppose $h^1, \ldots, h^t$ are a given set of scenarios with associated probabilities $f_1, \ldots, f_t$. Then as noted in Section 2, (1.1) can be put into the equivalent LP form :

minimize $cx + qy + f_1 qy^1 + \ldots + f_t qy^t$

subject to

$$Ax \qquad\qquad\qquad = b$$

$$Tx + Wy - \chi \qquad\qquad = 0$$

$$\chi + Wy^1 \qquad\qquad = h^1$$

$$\vdots \qquad\qquad \vdots$$

$$\chi \qquad +Wy^t = h^t$$

$$x, y, y^j \geq 0$$

(4.1)

Note that even in the above LP formulation it is worthwhile to make the problem transformation involving χ, since otherwise Tx would repeat itself in every row involving h^i. (4.1) is a much more sparse representation than the equivalent LP in which χ is not present. If there are relatively few scenarios, it would be practical to solve (4.1) directly. What is to be gained by a method based on GLP *even in this context?*

In the GLP approach, solving (1.5) (and in the process computing the objective row coefficients of (1.4)) can be the most taxing part of the computation. Under our present assumptions, this subproblem, namely

$$\underset{\chi \in \mathbb{R}^{m2}}{\text{minimize}} \; \Phi(\chi) \equiv \Psi(\chi) + \pi^K \chi \qquad\qquad (4.2)$$

can be expressed as:

$$\text{minimize} \; \pi^K \chi + f_1 q y^1 + \ldots + f_t q y^t$$

subject to

$$\chi + Wy^1 \qquad\qquad = h^1$$

$$\vdots \qquad\qquad \vdots$$

$$\chi \qquad + Wy^t = h^t$$

$$y^j \geq 0$$

(4.3)

Note that $\Phi(\chi)$ is polyhedral. Consider the following two ways of solving (4.2):

a) Use the revised simplex method to solve the equivalent LP problem
   (4.3) and take advantage of its very special structure. Note, in
   particular, that W occurs in each row but in different variables.
   This makes it likely that a feasible starting basis B can be found
   in variables $y^1, \ldots, y^t$ which is square-block diagonal with many submatrices on the diagonal repeating themselves. FTRAN and BTRAN operations can be done very efficiently with such a basis matrix, and subsequent iterations to find an optimal solution can be based on the
   Schur Complement Update, see Bisshop and Meeraus, 1977, and Gill et
   al., 1982, which retains the advantage of B.

b) Solve (4.2) using a minimization routine for non-smooth functions.
   Note, in particular, that the dimension of this problem is determined by the number of rows in the technology matrix T and this
   will often be small, even when the number t of realizations of the
   right-hand-side is large. An evaluation of $\phi(\chi)$ and its subgradient,
   say at the point $\bar{\chi}$, which will normally be required at each iteration of the minimizer, involves the solution of the following *se-*
   parable problem:

minimize $f_1 q y^1 + \ldots + f_1 q y^t$

subject to

$$Wy^1 \qquad = h^1 - \bar{\chi}$$
$$\vdots \qquad \vdots$$
$$Wy^t = h^t - \bar{\chi}$$

$$(4.4)$$

$$y^j \geq 0$$

and various techniques that go under the heading of *bunching* and
sifting, see Wets, 1983a, can now be profitably employed to substantially speedup the solution of (4.4). It is precisely these
techniques, coupled with the use of the dual simplex method which
give the L-shaped method for SLP, (see Birge, 1982), a substantial
edge over straight LP applied to (4.1). The same would hold true
for our method.

When t is large [*] we would not want to solve (1.4) unless a

---

[*] Suppose T had 10 rows, and the components $h_i(w)$ were independently
distributed, each with 3 possible levels. Then $t = 3^{10}$.

Schur Complement Update approach was attempted. Even then there might be difficulties, since n_1 could be large and consequently many columns of $\begin{pmatrix} A \\ T \end{pmatrix}$ could play a role in the optimal basis. In contrast, approaches based upon a) and b) above would still be viable. We have, for purposes of discussion, left χ unconstrained, and *minimized* $\Phi(\chi)$ in (4.2). In practice, there are *three* important points to note. First, not all elements of h(w) are necessarily stochastic. In this case the levels of the corresponding components of χ can be *fixed* in the solution of (4.3) as discussed in a) above, and in the solution of (4.4) as discussed in b). This reduces the dimensionality further. Recalling also the discussion after equation (1.5), we could restrict χ to the support of the distribution. This means we could often work with bound constrained problems of the form

minimize $\Psi(\chi) + \pi^K\chi$

subject to

$$\underline{l} \leq \chi \leq \underline{u} \tag{4.5}$$

with $l_i = u_i$ for some components. As an extreme case suppose only one element of h(w) in the recourse problem was stochastic; then (4.5) is, in effect, a unidimensional problem. The second point to note is that (4.2) does *not* have to be pushed to optimality. All we really need is a solution χ^{K+1} which satisfies $\Psi(\chi^{K+1}) + \pi^K\chi^{K+1} - \theta^K < 0$ where θ^K is the optimal dual multiplier on the convexity row of the master (1.3). This can easily be incorporated into the methods discussed above for solving the subproblem. Thirdly, it is likely that a good set of *initial* tenders can be specified, and this will again considerably speed up the convergence of the algorithm.

4.2. $\Psi(\chi)$ *must be approximated*

One approach is to use sampling and couple this with use of the stochastic quasi-gradient method (see Ermoliev, 1983) to solve the subproblem. Another approach is to proceed by repeated approximation of the distribution of h(w) and to compute bounds on $\Psi(\chi)$. Some preliminary suggestions are given in Birge, 1983. An important question is how to satisfactorily integrate the approximation strategy and the generalized programming algorithm, and the interpretation given in Theorem 2.2 may prove useful in this regard. We defer further discussion of this to a later date.

5. CONCLUSIONS

The methods introduced in this paper for solving SLP problems with recourse, involve the problem transformation (1.2), combined with the use of generalized linear programming. The problem transformation restricts the degree of nonlinearity to m_2, the number of rows of T and this, of course, enhances the efficiency of the GLP method. The problem transformation (1.2) is useful in other contexts. We have seen this already in (4.1) and the subsequent discussion. *We believe it could also be usefully employed within the L-shaped method,* see Van Slyke and Wets, 1969 and Birge, 1982, since each cut introduced would have at most m_2 elements rather than n_1, the dimension of x. For yet another example of such transformations, see Nazareth, 1983.

The approach discussed here could also be used to devise algorithms for solving a wider class of problems than (1.1). For example, cx, Ax-b = 0 and Tx could be replaced by nonlinear function c(x), $g(x) \leq 0$ and T(x) and a *nonlinear programming method* could then be used to solve the associated master. Also if T were *stochastic* we could apply GLP to (1.1), but now the degree of nonlinearity would be n_1. In practice only a few columns of T are normally stochastic. In this case, we could introduce a problem transformation $T_1 x_1 - \chi_1 = 0$ where T_1 represents the nonstochastic columns of T and x_1, the corresponding x-variables. Then GLP could be applied to a transformed problem whose degree of nonlinearity is only (number of stochastic columns of T) + (number of rows of T). Both these extensions deserve further exploration.

ACKNOWLEDGEMENT

I would like to thank Roger J-B Wets who introduced me to stochastic linear programming with recourse, and gave me the benefit of his deep insight into the theory of this subject. This made possible my work on algorithms for SLP problems discussed here. Many thanks also to the referee for some helpful comments and to Elfriede Herbst for typing this paper.

REFERENCES

[1]   Birge, J. (1982):Decomposition and Partitioning methods for
      multistage stochastic linear programs, Tech. Report 82-6,
      Dept. of IE & OR, University of Michigan.

[2]   Birge, J. (1983): Using sequential approximations in the
      L-shaped and generalized programming algorithms for sto-
      chastic linear programs, Tech. Report 83-12, Dept. of IE
      & OR, University of Michigan.

[3]   Bisschop, J. and Meeraus, A. (1977): Matrix augmentation and
      partitioning in the updating of the basis inverse, *Mathe-
      matical Programming*, 18, 7-15.

[4]   Cleef, H. (1981): A solution procedure for the two-stage
      stochastic program with simple recourse, *Z. Operations
      Research*, 25, p. 1-13.

[5]   Dantzig , G.B. (1963): *Linear Programming and Extensions*,
      Princeton University Press.

[6]   Eaves, B.C. and Zangwill, W. (1971): Generalized cutting plane
      algorithms, *SIAM J. Control*, 9, p. 529-542.

[7]   Ermoliev, Yu. (1983): Stochastic quasigradient methods and
      their application in systems optimization, *Stochastics*,
      9, p. 1-36.

[8]   Gill, P.E., Murray, W., Saunders, M.A. and Wright, M.H. (1982):
      Sparse matrix methods in optimization, Tech. Report SOL-82-
      17, Systems Optimization Lab., Dept. of Operations Research,
      Stanford University.

[9]   Ho, J. (1974): Nested decomposition for large scale linear
      programs with the staircase structure, Report SOL-74-4.
      Systems Optimization Lab., Dept. of Operations Research,
      Stanford University.

[10] Kallberg, J. and Kusy, M. (1976): Code Instruction for S.L.P.R.,
      a stochastic linear program with simple recourse, Tech.
      Report, University of British Columbia.

[11] Lemarechal, C. (1982): Numerical experiments in nonsmooth
      optimization, In: *Progress in Nondifferentiable Optimization*,
      E.A. Nurminski (Ed.), IIASA Collaborative Proceedings Series
      CP-82-S8, p. 61-84.

[12] Murtagh, B. and Saunders, M. (1978): Large-scale linearly con-
      strained optimization, *Mathematical Programming* 14,
      p. 41-72.

[13] Nazareth, L. (1983): Variants on Dantzig-Wolfe decomposition
      with applications to multistage problems, IIASA Working
      Paper, WP-83-61, Laxenburg, Austria.

[14] Nazareth, L. and Wets, R. J-B. (1983): Algorithms for stochastic
      programs: the case of nonstochastic tenders, IIASA Working
      Paper, WP-83-5 (revised version to appear in forthcoming
      Mathematical Programming Study).

[15] Nazareth, L and Wets, R.J-B. (1984): Stochastic programming
      with recourse: algorithms and implementation, IIASA Work-
      ing Paper (forthcoming).

[16] Parikh, S.C. (1968): Lecture notes on stochastic programming,
      unpublished, University of California, Berkeley.

[17] Polyak, B. (1978): Nonlinear programming methods in the pre-
      sence of noise, *Mathematical Programming*, 14, p. 87-97.

[18] Shapiro, J.F. (1979): Mathematical Programming: Structures and
      Algorithms, John Wiley, New York.

[19] Van Slyke, R. and Wets, R. (1979): L-shaped linear programs with
      applications to optimal control and stochastic linear
      programs, *SIAM J. on Appl. Math.*, 17, pp. 638-663.

[20] Wets, R. (1972): Characterization theorems for stochastic pro-
      grams, *Mathematical Programming*, 2, 166-175.

[21] Wets, R. (1974): Stochastic programming, unpublished, Lecture
      Notes, University of California, Berkeley.

[22] Wets, R. (1983a): Stochastic programming: approximation schemes
      and solution techniques, In: *Mathematical Programming
      1982: The State-of-the-Art*, Springer-Verlag, Berlin.

[23] Wets, R. (1983b): Solving stochastic programs with simple re-
      course, *Stochastics*, 10, p. 219-242.

[24] Williams, A.C. (1966): Approximation formulas for stochastic
      linear programming, *SIAM J. Appl. Math.*, 14, No. 4,
      p. 668-677.

[25] Ziemba, W.T. (1972): Solving nonlinear problems with stochastic
      objective functions, *Journal of Financial and Quantitative
      Analysis*, VII, p. 1809-1827.

ON THE USE OF NESTED DECOMPOSITION FOR SOLVING

NONLINEAR MULTISTAGE STOCHASTIC PROGRAMS

Marie Cécile Noël and Yves Smeers

CORE

1. INTRODUCTION

Nested decomposition was initially proposed by Glassey (1973) for handling staircase linear programming problems. An indepth treatment of the
method with numerical experiments can be found in Ho (1978) and Ho and Manne
(1974). A most common interpretation of staircase linear programs is to see
them as multistage deterministic decision problems : variables are associated with different steps of the decision process and it is assumed that action
taken at some stage influence the set of possible decisions at the next one.
The matrix of the corresponding linear programming problem has the form given
in figure 1.

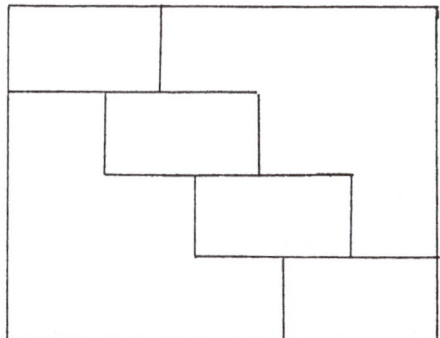

Figure 1 : *block structure of a staircase linear programming problem.*

The initial nested decomposition algorithm has been extended into several directions. Ament et al (1981) consider the more general problem where
decision taken at some step may influence the set of possible decisions in
several successive stages : the matrix of the problem has now a lower block
triangular structure (figure 2). They also describe an implementation of
the method based on MPSX/370 and report on the solution of large test problems. Nested decomposition has been extended to nonlinear multistage prob-

lems with staircase structure (in this case the Jacobian of the constraint
matrix has the staircase structure) (O'Neill (1976)) but without any numerical experiment.

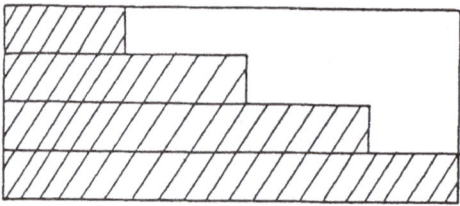

Figure 2 : lower block triangular matrix.

Decomposition was early recognized as a tool for handling two stage stochastic problems (Van Slijke and Wetz (1969)). More recently, Birge (1982)
has shown how the dual of the nested decomposition algorithm could be applied
to solve multistage discrete stochastic problems; he also proposes improvements
of the methods based on a partitionning of the feasible set at each stage.
Related work can be found for the deterministic multistage problem in
Abrahamson (1980) and Wittrock (1983).

Quadratic programming constitutes the only case of nonlinear model for
which stochastic programming methods have been proposed and implemented
(Louveaux (1978) and (1980)). We deal in this paper with nonlinear multistage
problems where decisions in some stage influence the feasible set in all possible successive stages. With respect to the existing literature our problem
can thus be considered as an adaptation to multistage discrete stochastic
models of a nonlinear (O'Neill (1976)) version of the algorithm designed by
Ament et al (1981) for lower block triangular problems. Any discussion of
nested decomposition requires tedious notations. In order to shorten the pre-
sentation we skip the subject altogether and send the reader to the existing
literature. The application of the method is illustrated on a simple economic
growth model introduced in section 2. Implementing a nonlinear version of
nested decomposition requires a robust nonlinear programming code. Although
this is by no mean necessary for handling the small test problems considered
here we based our coding on MINOS (Murtagh, Saunders (1977)) that we use essentially as a block box. Even so, a "quick and dirty" implementation of the
algorithm raises several questions that are briefly surveyed in section 3.
Finally numerical results are given in section 4.

2. TEST PROBLEMS

We consider a stochastic version of the following simple growth model. Consider an horizon of n periods over which we want to maximize some intertemporal utility function which depends on private consumption in each period. The problem is formulated as follows. We assume that the output of the economy in each period i is given by a four factor production function

$$F_i(K_i,L_i,E_i,N_i) = \left\{ a(E_i^\beta N^{1-\beta})^\rho + b(K_i^\alpha \bar{L}^{1-\alpha})^\rho \right\}^{1/\rho}$$

where K_i, L_i, E_i and N_i respectively denote the capital, labour, electrical and nonelectrical energy available in period i. This output is allocated among private consumptions (C_i), investments (I_i) and payments for electrical $(p_{e_i} E_i)$ and nonelectrical energy $(p_{n_i} N_i)$

$$C_i + I_i + p_{e_i} E_i + p_{n_i} N_i - F(K_i,L_i,E_i,N_i) \leqslant 0 \, ,$$

where we assume in this equation that the amount of labour available in i is given.

The model and the associated data are extracted from the version of ETA-MACRO (Manne (1977) and Manne et al (1983)) established by Manne and Sira (1980) for the European countries of O.E.C.D. Coming to the construction of the stochastic model, we decompose the horizon into five year horizon periods. We specify the extensive form of the model by considering a binary event tree rooted in year 80. A branch coming out of a node corresponds to an evolution of the price of fossil energy during the period. We deal with two scenarios : in the first one the price escalations are respectively 1 and 3 % while there are 0 and 4 % in the second one. Each branch from a node is assigned a 50 % probability. The median of the price evolution is thus 2 % which is compatible with the findings of the International Energy Workshop (Manne and Schrattenholtzer (1983)).

In order to introduce a valuation of the residual capital stock in the objective function we complete the binary tree by adding a branch from each terminal node : the number of periods considered in the event tree then becomes n+1 . All investment decisions are restricted however to the first n periods. We let T_{n+1} denote this extended tree and L_{n+1} be the set of its terminal nodes; N is the number of nodes in the binary tree, and $p(i)$ refers to the predecessor of i in T_{n+1} .

We can state the extensive form of our stochastic program as

$$\text{Max} \sum_{i=1}^{N} \alpha_i \log C_i + \sum_{i \in L_{n+1}} \alpha_i K_i$$

s.t.
$$C_i + I_i + p_{e_i} E_i + p_{n_i} N_i - F(K_i, \overline{L}_i, E_i, N_i) \leq 0 , \quad i = 1, \ldots, N \tag{1}$$

$$K_i - \lambda K_{p(i)} - \xi I_{p(i)} \leq 0 , \quad i \in T_{n+1} , \quad i \neq 1 \tag{2}$$

$$K_1 \quad \text{given} ,$$

where – the terms $\alpha_i K_i$ are introduced for dealing with horizon effects

– λ denotes the decay factor of a unitary capital stock in a period on the planning horizon,

– ξ is a coefficient that expresses the capital stock available in a given period resulting from a unitary investment realized throughout the period before of the node in the tree.

– α_i is the product of the discount factor at node i by the probability of the node in the tree.

Let u_{i1} and u_{i2} be the dual variables associated with the constraints (1) and (2) respectively. In order to write the dual of our problem, we introduce the function $g_i(u_{i1}, d_i)$ which is equal to

$$\max_{K_i, E_i, N_i} u_{i1} \left[\frac{d_i}{u_{i1}} K_i + p_{e_i} E_i + p_{n_i} N_i - F(K_i, \overline{L}_i, E_i, N_i) \right]$$

when i is not the root, and to

$$u_{11} \left[\max_{E_1, N_1} (p_{e_1} E_1 + p_{n_1} N_1 - F(\overline{K}_1, \overline{L}_1, E_1, N_1)) \right]$$

when i denotes the root.

The dual problem can then be stated as

$$\text{Max} \sum_{i=1}^{N} \left[\alpha_i \left(1 + \log \frac{u_{i1}}{\alpha_i} \right) + g_i(u_{i1}, d_i) \right] - \lambda \overline{K}_1 \left(\sum_{i \in S(1)} u_{i2} \right) \tag{3}$$

s.t.
$$u_{i1} - \xi \sum_{j \in S(i)} u_{j2} \geq 0 \quad i \in T_{n+1} \tag{4}$$

$$d_i = u_{i2} - \lambda \sum_{j \in S(i)} u_{j2} \qquad i \in T_{n+1} \quad , \quad i \neq 1 \tag{5}$$

$$u_{i2} \geqslant \alpha_i \qquad i \in L_{n+1} \quad . \tag{6}$$

It is on this dual that we shall apply the nested decomposition algorithm. The test problems have been constructed for models of 5, 7 and 9 periods and for binary event trees. The sizes of the primal and dual problems are given in Table 1.

Primal problem	Number of constraints		Number of variables
	linear	nonlinear	
5 periods	46	31	170
7 periods	190	127	698
9 periods	766	511	2 810

Dual problem	Number of constraints	Number of variables	
		linear	nonlinear
5 periods	61	46	61
7 periods	253	190	253
9 periods	1 021	766	1 021

Table 1 : Statistics of the different problems

3.   IMPLEMENTATION

The object of this paper is to provide a first investigation of the use of nested decomposition for solving some stochastic economic growth problems : the implementation of the method reported here is thus rather elementary and does not aim at efficiency; it however already raises various questions that need to be handled in any further implementation.

Different approaches can be followed for implementing the nested decomposition algorithm. In order to characterize our coding we shall consider the two following design philosophies :

i) The optimization code (in this case MINOS) is used as a black box and information between subproblems is transferred through files (in this case INPUTMINOS, SPECMINOS, BASISMINOS, SCRATCHMINOS and LISTMINOS).

ii) Transfer of information between the different subproblems is performed through the internal data structure of the optimizer. It is organized using algorithmic facilities (see Ho and Loute (1981)) so as to maximize efficiency.

The first approach is clearly the most straightforward from a conceptual point of view. It can reasonably be implemented for algorithms that do not require to iterate between many subproblems (see Ginsburgh and Waelbroeck (1980)) for an example of such an application in the field of general equilibrium). It is clearly infeasible here where a single cycle of the algorithm for the nine periods problem would require to open and close more than 2000 files (5 files for each call to MINOS and 511 problems to handle).

The second approach is guaranteed to maximize efficiency : it however requires a deep understanding of the data structure used by the algorithm and is not easily implemented when the optimization code has been designed so as to offer access to computation tools and parts of internal data structure. MINOS does not need this last requirement and an efficient coding of the method would have proved prohibitive given the exploratory nature of this work. A medium approach has thus been followed : the input and output functions of MINOS have been modified so as to eliminate the opening and closing of all the files associated with each problem : the optimizer has been used as a black box but most of the data transfer has been done through vectors instead of files. We now proceed to describe this implementation in more detail, starting from the set of MPS inputs constructed by the matrix generator (OMNI/PDS (1980)).

The first step of the procedure reads the sequence of MPS inputs and reorganizes them in a single direct access file. Three operations are performed during this phase : the problems are reorganized according to the sequence in which they will be considered by the decomposition algorithm. This is done using a table of correspondence provided by the user (it is thus required that the user knows the sequence adopted by the algorithm). The row and rhs sections of each problem are modified by the addition of a convexity row; finally the column section is expanded so as to allow for the handling of proposals and of the objective function. These transformed problems are then copied in a direct access file.

In order to avoid the opening and closing of too many files, the specification, input and basis files of the different problems have all been copied in three files and the input routine of MINOS modified to allow a direct access of these latters. The information relative to a given problem is also packed in the smallest possible number of records to decrease input time : finally proposals are directly written in that file in order to avoid the use of a proposal file. This organization is illustrated on Figure 3.

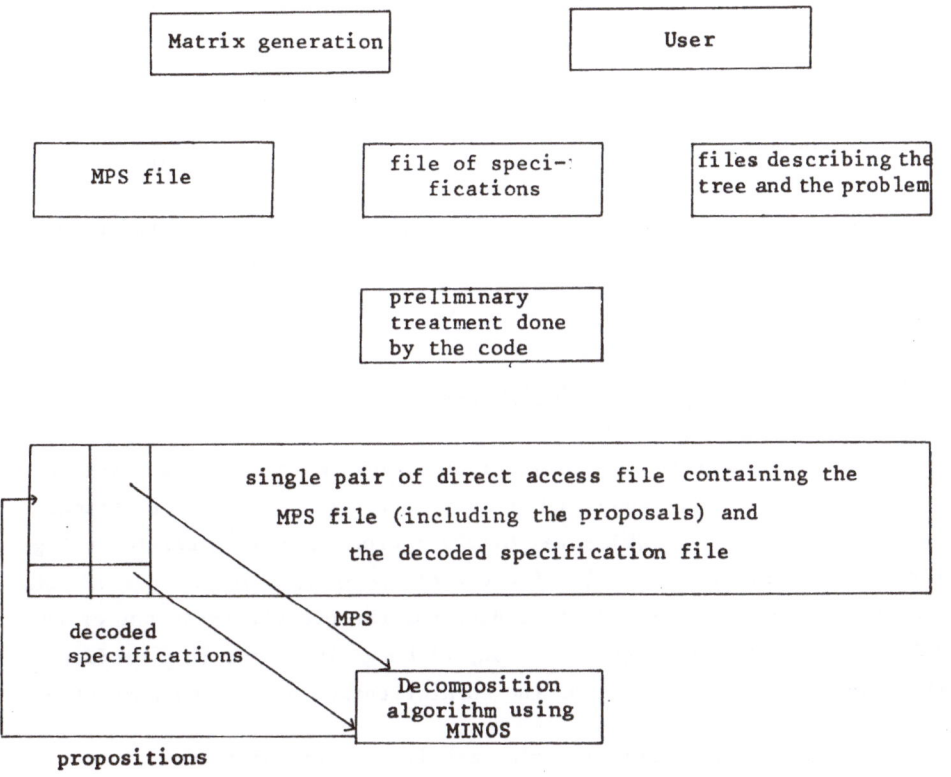

Figure 3 : Using MINOS with modified input-output

We conclude this section by noting some shortcomings that will have to be overcome in future implementations.

Transfer of information between the subproblems

In the current implementation, proposals of a subproblem are constructed by matrix multiplication of the optimal solution (or ray) of an other sub-problem. Similarly the modification of the objective function is made by simply subtracting from the linear part of the objective function a vector obtained by postmultiplying the dual solution of a subproblem by the matrix of coupling constraints. Ho and Loute (1981) use data structures that allow these operations to be performed much more efficiently.

Treatment of the problems at each cycle

Rows and columns names are decoded at every cycle for each subproblem. This is clearly unnecessary and could be avoided by storing the correspondence between internal and external codes as obtained at the first cycle.

Needless to say implementing such facilities would require the coding of additional interface routines to accomodate the direct access nature of the files now used by the algorithm.

4. NUMERICAL RESULTS AND CONCLUSIONS

This section presents some of the results of the experiments using our code on the test problems described in section 2. All runs were performed under AOS/VS on a DATA GENERAL model MV8000 : problems were generated using OMNI/PDS on an IBM/370 model 158. Because of the exploratory nature of this work we only report here results that describe the overall behaviour of the algorithm and leave aside information that, although important in a production environment, essentially depends on the sophistication of the implementation.

Table 2 gives the number of cycles required by the algorithm.

	5 periods	7 periods	9 periods
1 % - 3 %	12	22	27
0 % - 4 %	12	21	28

Table 2 : Number of cycles for achieving optimality.

Optimality is claimed when the subproblems stop generating proposals (or do not iterate on generate proposals) or when the ratio

current objective function value - best dual solution found so far
----
current objective function value

becomes smaller than 10^{-6} . It is interesting to note that the number of cycles seems to increase linearly with the number of periods and remains in any case quite reasonable.

Table 3 gives the evolution of the objective function value and consumption for the last cycles of the algorithm.

A good estimate of the objective function is reached quite quickly with most of the remaining cycles resulting in only small improvements. In contrast, all the cycles are necessary in order to achieve the convergence of the consumption and investment variables. In that sense it does not seem reasonable to cut the progress of the algorithm too quickly.

Nested decomposition appears, on the basis of this limited experience, to be a reasonable approach for handling multistage stochastic problems. Needless to say, a full scale implementation of the method would require much more effort than the one presented here.

Scenario \\ Cycles		-5	-4	-3	-2	-1	0
5 periods (increase of P_n : 0-4 %)	o	19856.882	19834.837	19794.317	19793.796	19792.678	19792.678
	c	2406.763	2233.031	2253.980	2335.138	2302.329	2302.329
5 periods (increase of P_n : 1-3 %)	o	19856.997	19834.980	19794.440	19793.911	19792.796	19792.796
	c	2406.762	2233.983	2253.247	2336.330	2302.010	2302.010
7 periods (increase of P_n : 0-4 %)	o	22896.156	22890.556	22890.326	22890.306	22890.123	22890.027
	c	2346.438	2278.434	2301.355	2318.201	2293.251	2315.720
7 periods (increase of P_n : 1-3 %)	o	22890.820	22890.729	22890.468	22890.317	22850.28	22890.259
	c	2292.593	2309.334	2315.602	2296.203	2316.384	2305.680
9 periods (increase of P_n : 0-4 %)	o	24781.220	24780.716	24780.345	24780.211	24780.162	24780.028
	c	2312.260	2332.482	2289.085	2311.161	2311.480	2311.161
9 periods (increase of P_n : 1-3 %)	o	24781.566	24780.979	24780.763	24780.681	24780.375	24780.336
	c	2315.243	2328.316	2309.913	2297.066	2306.528	2321.032

Table 3 : *Evolution of the objective value and consumption.*

o and c denote respectively the objective value and the consumption (in 10^9 \$).

REFERENCES

Abrahamson, P.G. "A Nested Decomposition Approach for Solving Staircase in: *Large-Scale Linear Programming*. Ed. G.B. Dantzig, M.A.H. Dempster and M.J. Kallio, IIASA CP-81-S1, pp. 367-381.

Ament, D., Ho, J., Loute, E. and Remmelswall,M., "LIFT : A Nested Decomposition Algorithm for Solving Lower Block Triangular Linear Programs", in: *Large-Scale Linear Programming*. Ed G.B. Dantzig, M.A.H. Demptster and M.J. Kallio, IIASA, CP-81-S1, pp. 383-408.

Birge, J.R., "Decomposition and Partitioning Methods for Multi-Stage Stochastic Linear Programs", Technical Report n° 82-6, University of Michigan, May 1982.

Ginsburgh, V. and J. Waelbroeck, *Activity Analysis and General Equilibrium Modelling*, Amsterdam, North-Holland, 1981.

Glassey, C.R., "Nested Decomposition and Multi-Stage Linear Programs", *Management Science*, 20, pp. 282-292, 1973.

Ho, J.K. and A.S. Manne, "Nested Decomposition for Dynamic Models", *Mathematical Programming*, 6, 121-140, 1974.

Ho, J.K., "Implementation and Application of a Nested Decomposition Algorithm", in:*Computers and Mathematical Programming*, W.W. White (ed.), pp. 21-30, NBS, 1978.

Ho, J.K. and E. Loute, "An Advanced Implementation of the Dantzig-Wolfe Decomposition Algorithm for Linear Programming", *Mathematical Programming*, 20, pp. 303-326, 1981.

Louveaux, F.V., "Piecewise Convex Programs", *Mathematical Programming*, 15, pp. 53-62, 1978.

Louveaux, F.V., "A Solution Method for Multistage Stochastic Programs With Recourse with Application to an Energy Investment Problem", *Operations Research*, 28(4), pp. 889-902, 1980.

Manne, A.S., "ETA-MACRO : A Model of Energy-Economy Interactions", Research Project 1014, Department of Operations Research, Stanford University, Stanford, California 94305, December 1977.

Manne, A.S. and T. Sira, "European Energy Supplies and Demands : A Long-Term Perspective", International Energy Program, Discussion Paper, Stanford University, Stanford, California 94305, October 1980.

Manne, A.S. and L. Schrattenholtzer, "A Summary of the 1983 Poll Responses", International Energy Workshop, IIASA, Laxenburg, Austria, June 1983.

Manne, A.S., Beltrano, M.A., Rutherford, T.F., Svoronos, A.N. and T.F. Wilson, "ETA-MACRO : A Progress Report", Research Project 1014, Department of Operations Research, Stanford University, Stanford, California 94305, July 1983.

Murtag, A.B. and M.A. Saunders, "MINOS : A Large-Scale Nonlinear Programming System", User's guide, Technical Report SOL 77-9, 1977.

OMNI Linear Programming System, Version 4.0, Haverly System Inc., July 1980.

O'Neill, R.P., "Nested Decomposition of Multistage Convex Programs", *SIAM Journal on Control and Optimization*, 14(3), pp. 409-418, may 1976.

Van Slijke, R. and R. Wets, "L-Shaped Linear Programs with Applications to Optimal Control and Stochastic Linear Programs", *SIAM Journal on Applied Mathematics*, 17, pp. 638-663, 1969.

Wittrock, R.J., "Advances in A Nested Decomposition Algorithm for Solving Staircase Linear Programs", Draft, 1983.

CONTRIBUTIONS TO THE METHODOLOGY OF
STOCHASTIC OPTIMIZATION

János Pintér

Research Center for Water Resources Development

H-1453 Budapest, P.O.Box 27. Hungary

1. INTRODUCTION

In the practice of engineering and economic design, the decisions to be accepted frequently depend on random factors in a complicated manner. Water resources management problems (with partly unknown climatic and hydrological circumstances, physical and biochemical processes, varying water demands and treatment technologies, etc.) may serve as typical examples of this situation (cf. e.g. [4,8,11,18,19,21]); one can also mention decision problems connected with industrial process design, mass service systems, multi-item stock planning, etc.. These decision problems often can be formulated as stochastic programming models (cf. e.g. [5,7,17,23]):

$$\min E[f(x,y)]$$

$$E[g_i(x,y)] \geq 0, \quad i = 1,\ldots,m \qquad (1.1)$$

$$x \in K.$$

In (1.1) x is a decision variable, represented as a point from some subset K of the n-dimensional real Euclidean space R^n, $y = y(\omega)$ is a q-dimensional vector-valued random variable, f, $g_i : R^{n+q} \rightarrow R^1$, $i = 1,\ldots,m$ are Borel-measurable functions and E is the symbol of mathematical expectation (with respect to y).

Generally speaking, it is an essential feature of stochastic optimization problems of the form (1.1) that the values of the figuring functions are computable only in presence of some error (random noise), while to increase the evaluation accuracy more resources (experiments, simulation cycles, etc.) are necessary. Moreover, convexity and/or differentiability properties of the problem in question may not hold or may be unknown. Under these circumstances an optimization strategy should define a way for producing a sequence of improving decisions which are evaluated with gradually increasing accuracy.

In the present paper both mentioned aspects of stochastic optimization strategies are briefly dealt with. In Section 2 a general framework for constructing stochastically combined (hybrid) optimization algorithms is outlined: these results provide a broad class of algorithms which are globally convergent also in lack of convexity and smoothness properties, while their (local) efficiency is usually superior to that of some other global methods. In Section 3 a modified Bernstein-technique is presented for estimating noise-perturbed fuction values.

2. STOCHASTICALLY COMBINED OPTIMIZATION ALGORITHMS

As it is well-known, in principle there exist different methods which are globally convergent under mild analytical assumptions about the structure of a problem such as e.g. (1.1). These methods are often of simultaneous (passive) nature (e.g. grid search or pure random search), and usually become rather inefficient in the neighbourhood of (local or global) optima, where they apparently should be "switched over" to some locally more efficient procedure (cf. e.g. [2,3,9,11,12, 22]). Based on this heuristic idea, in [14] a general class of stochastically combined optimization procedures was defined. Here only the main ideas are briefly presented: more details can be found in [13, 16].

Consider the [deterministic] global optimization problem

$$\min_{x \in C} f(x), \qquad (2.1)$$

where $f: R^n \to R^1$ is a Borel-measurable function, bounded from below on the Borel-set $C \subset R^n$. Assume that the set of optimal solutions of (2.1) is nonempty: denote this set by C^*; moreover, define $f^* = f(x^*)$ for $x^* \in C^*$.

To find points of C^*, a stochastic sequence $\{x_k\}$ is generated according to the rule

$$x_{k+1} = D_k(\bar{z}_k, \bar{\omega}_k) \in C, \qquad k = 0,1,2,\ldots \qquad (x_o \in C)$$

$$\qquad (2.2)$$

$$\bar{z}_k = (x_o, f(x_o), x_1, f(x_1), \ldots, x_k, f(x_k)) \qquad \bar{\omega}_k = (\omega_o, \omega_1, \ldots, \omega_k).$$

This means that x_{k+1} is selected according to some conditional probability measure μ_k given on the measurable space $(C, \mathcal{B}_C)$ ($\mathcal{B}_C$ is the set of Borel-subsets of C). In other words, the decision rule D_k depends on the search information $\bar{z}_k$ and the k-th projection of the sequence of random factors $\omega = (\omega_0, \omega_1, \omega_2, \ldots)$ influencing the decisions D_k $k = 0,1,2,\ldots$. Let the realizations of ω belong to the set Ω: we shall suppose that to the above decision scheme there corresponds a probability measure P defined on the measurable space (Ω, A). We want to find sufficient conditions under which for the sequence $\{x_k\}$ there holds

$$P(\lim_{k \to \infty} \rho(x_k, C^*) = 0) = 1 \qquad \rho(x, C^*) = \inf_{x^* \in C^*} \| x - x^* \|.$$

In a stochastically combined way, denoted symbolically as

$$D_k : \alpha_k G_k + (1-\alpha_k) L_k. \tag{2.3}$$

This means that - independently of $\bar{z}_k$ and $\bar{\omega}_k$ (i.e. the results of the previous iteration steps) - at step k subprocedure G_k is applied with probability α_k, while subprocedure L_k is applied with probability $(1-\alpha_k)$. (From the result to be presented below it becomes obvious that the sequence $\{G_k\}$ assures theoretical convergence properties of the algorithm scheme (2.2)-(2.3), while the sequence $\{L_k\}$ can be chosen in principle arbitrarily (in order to enhance efficiency)).

In order to solve (2.1), let us specify the rule (2.2) as follows:

$$x_{k+1} = y = D_k(\bar{z}_k, \bar{\omega}_k) \in C \qquad k = 0,1,2,\ldots$$

is accepted *iff* $f(y) < f(x_k)$; otherwise set $x_{k+1} = x_k$.

Define for $\hat{\delta} > 0$ and $\hat{\varepsilon} > 0$ the sets

$$G(C^*, \hat{\delta}) = \{y \in C : \rho(y, C^*) < \hat{\delta}\} \qquad \hat{C} = \hat{C}(\hat{\delta}) = C \setminus G(C^*, \hat{\delta})$$

and the random events

$$F_k = F_k(\hat{\delta}) : \{x_k \in \hat{C}\} , \quad A_k = A_k(\hat{\varepsilon}) : \{f(x_{k+1}) \leq f(x_k) - \hat{\varepsilon}\}, \quad k = 0,1,2,\ldots$$

For arbitrary random events $A, B \in A$ define $P(A|B) = P(A)$ if $P(B) = 0$; moreover, let us denote by $P(A|B; G_k)$ and $P(A|B; L_k)$ the conditional probability of A, given B *and* the application of subprocedure G_k or L_k, respectively.

THEOREM 2.1. If for arbitrary $\hat{\delta} > 0$ there exist $\hat{\varepsilon} = \hat{\varepsilon}(\hat{\delta}) > 0$ and $\{\hat{p}_k\}$, $0 \leq \hat{p}_k = \hat{p}_k(\hat{\delta}) \leq 1$, $k = 0,1,2,\ldots$ such that for every pair of indices k, n, $k \geq n$ the relations

$$P(A_k \cup \bar{F}_k | \bigcap_{j=n}^{k-1} (\bar{A}_j \cap F_j); G_k) \geq \hat{p}_k , \quad \sum_{k=0}^{\infty} \alpha_k \hat{p}_k = \infty \tag{2.4}$$

are valid, then

$$P(\liminf_{k \to \infty} \rho(x_k, C^*) = 0) = 1, \qquad (2.5)$$

i.e. there exists (at least) a subsequence of $\{x_k\}$ which converges to the set of optimal solutions of (2.1), w.p. 1.

PROOF. Define the random event

$$F = F(\delta) : \{\liminf_{k \to \infty} \rho(x_k, C^*) \geq \delta\};$$

evidently, it is sufficient to show that for arbitrary $\delta > 0$ there holds $P(F(\delta)) = 0$. Assume indirectly that for some $\delta > 0$ $P(F(\delta)) > 0$. As for $\omega \in F(\delta)$ we have $\rho(x_k, C^*) \geq \frac{\delta}{2}$ if $k \geq n = n(\delta, \omega)$, hence

$$F(\delta) \subseteq \bigcup_{n=0}^{\infty} \bigcap_{k=n}^{\infty} F_k(\frac{\delta}{2}).$$

Let $\hat{\delta} = \frac{\delta}{2}$, then by (2.4) we have that

$$P(\bigcap_{k=n}^{N} (\bar{A}_k \cap F_k)) = \prod_{k=n}^{N} P(\bar{A}_k \cap F_k | \bigcap_{j=n}^{k-1} (\bar{A}_j \cap F_j)) = \prod_{k=n}^{N} [1 - P(A_k \cup \bar{F}_k | \bigcap_{j=n}^{k-1} (\bar{A}_j \cap F_j))]$$

$$\leq \prod_{k=n}^{N} [1 - \alpha_k P(A_k \cup \bar{F}_k | \bigcap_{j=n}^{k-1} (\bar{A}_j \cap F_j); G_k)] \leq \prod_{k=n}^{N} (1 - \alpha_k \hat{p}_k) \leq \exp(-\sum_{k=n}^{N} \alpha_k \hat{p}_k) \xrightarrow[N \to \infty]{} 0, \quad n = 0, 1, 2, \ldots$$

Define $A_\infty = \bigcap_{n=0}^{\infty} \bigcup_{k=n}^{\infty} A_k$, then the above chain of relations implies that

$$P(\bar{A}_\infty \cap F) \leq P((\bigcup_{n=0}^{\infty} \bigcap_{k=n}^{\infty} \bar{A}_k) \cap (\bigcup_{m=0}^{\infty} \bigcap_{k=m}^{\infty} F_k)) \leq \sum_{n=0}^{\infty} \sum_{m=0}^{\infty} P((\bigcap_{k=n}^{\infty} \bar{A}_k) \cap (\bigcap_{k=m}^{\infty} F_k)) = 0$$

(as $P(\bigcap_{k=n}^{\infty} (\bar{A}_k \cap F_k)) = 0$ implies that $P((\bigcap_{k=n}^{\infty} \bar{A}_k) \cap (\bigcap_{k=m}^{\infty} F_k)) = 0$

$$n, m = 0, 1, 2, \ldots).$$

As hence evidently follows $P(A_\infty | F) = 1$, by the definition of the events $\{A_k\}$ we can conclude that

$$P(\lim_{k \to \infty} f(x_k) = -\infty | F) = 1,$$

i.e.

$$P(\lim_{k \to \infty} f(x_k) = -\infty) \geq P(F(\delta)) > 0,$$

while the last relation contradicts to the assumption that f is bound-ed from below on C. Therefore $P(F(\delta)) = 0$ for each $\delta > 0$, i.e. the assertion of the Theorem is valid.

REMARK 2.1. Assumption (2.4) means, roughly speaking, that the subprocedures $\{G_k\}$ are able to find improving feasible solutions $\{x_k\}$ with "sufficiently high" probabilities. It is easy to see that e.g. many adaptive random search strategies $\{G_k\}$ satisfy (2.4) (cf. e.g. [13,20]).

COROLLARY 2.1. If - besides the assumptions of Theorem 2.1. - we suppose that f is uniformly continuous on C, and for arbitrary $\delta > 0$ there exists $\varepsilon = \varepsilon(\delta) > 0$ such that

$$x \in C \text{ and } \rho(x,C^*) \geq \delta \text{ imply that } f(x) \geq f^* + \varepsilon, \qquad (2.6)$$

then

$$P(\lim_{k \to \infty} \rho(x_k,C^*) = 0) = P(\lim_{k \to \infty} f(x_k) = f^*) = 1.$$

PROOF. From uniform continuity of f and convergence of $\{f(x_k)\}$ there follows by (2.5) that

$$P(\lim_{k \to \infty} f(x_k) = f^*) = P(\exists\{k_i(\omega)\} \subseteq \{k\} : \lim_{i \to \infty} f(x_{k_i}) = f^*)$$

$$\qquad (2.7)$$

$$= P(\exists\{k_i(\omega)\} \subseteq \{k\} : \lim_{i \to \infty} \rho(x_{k_i},C^*) = 0) = P(\lim_{k \to \infty} \inf \rho(x_k,C^*) = 0) = 1.$$

Define now for arbitrary $\theta > 0$ the random event

$$K(\theta) : \{\lim_{k \to \infty} \sup \rho(x_k,C^*) \geq \theta\}.$$

Let $\delta = \frac{\theta}{2}$, then by (2.6) for $\varepsilon = \varepsilon(\delta)$ the following chain of rela-tions is valid:

$$K(\theta) \subseteq \bigcap_{n=0}^{\infty} \bigcup_{k=n}^{\infty} \{\rho(x_k,C^*) \geq \delta\} \subseteq \bigcap_{n=0}^{\infty} \bigcup_{k=n}^{\infty} \{f(x_k) \geq \varepsilon\} \subseteq \{\lim_{k \to \infty} f(x_k) \geq f^* + \varepsilon\}.$$

Hence, by (2.7) $P(K(\theta)) = 0$ for $\theta > 0$. This completes the proof.

Based on analogous argumentations to those of Theorem 2.1 and its Corollary, fairly general global or local convergence properties of different stochastic optimization strategies can be proved, covering also a class of noise-perturbed optimization problems: for more details the reader is referred to [13,14,16]. The computational experiments with adaptive stochastic local and global algorithms, described in [12,13], illustrate the practical applicability of the presented general framework.

3. ESTIMATION OF NOISE-PERTURBED FUNCTION VALUES

Assume now that (e.g. in the course of solving a stochastic optimization problem) the values of some (bounded) function

$$H(x) = Eh(x,y) \tag{3.1}$$

are not analytically computable, but are to be estimated (for different arguments $x \in R^n$). This estimation is based on the realizations of the independent and identically to y distributed (i.i.d.) random variables (r.v'.s) $y_k = y_k(\omega) \in R^q$, $k = 1,2,3,\ldots$ (h:$R^{n+q} \to R^1$ is a Borel-measurable function). According to the introductory remarks of Section 1, in such cases it is desirable to decrease the number of necessary realizations of the involved random factors as much as possible, while prescribing accuracy and reliability levels (confidence intervals) for the estimated value (3.1). In other words, one is interested to find a possibly close estimate of the minimal n, for which there holds

$$P(|\frac{1}{n} \sum_{k=1}^{n} h(x,y_k) - H(x)| \geq \varepsilon) \leq \delta \qquad (\varepsilon > 0, \quad 0 < \delta < 1). \tag{3.2}$$

For an arbitrary r.v. ξ with expectation $E(\xi)$ and variance $D^2(\xi)$, by the Bienaymé-Chebyshev inequality we have

$$P(|\xi - E(\xi)| \geq \lambda D(\xi)) \leq \frac{1}{\lambda^2} \qquad (\lambda > 1). \tag{3.3}$$

Based on the observation that for the r.v. $\zeta = e^{v(\xi - E(\xi))}$ $(v > 0)$ the Markov-inequality yields

$$P(\xi \geq E(\xi) + \frac{t + \ln E[e^{v(\xi - E(\xi))}]}{v}) \leq e^{-t} \qquad (t > 0), \tag{3.4}$$

in [1] it was proved that for bounded summands (3.3) can be specified in a significantly sharpened form. In [6] an overview is presented on several improvements of (3.3). Here a Bernstein-type improvement of (3.3) is given for estimating noise-perturbed function values: in the special case of relative frequency estimates of probabilities, this technique yields (essentially) the result of [10].

Define the r.v.'s and respective probability distribution functions (p.d.f.'s) as follows:

$$\xi_k^*(x) = h(x, y_k) , \qquad G_x^*(z) = P(\xi_k^*(x) < z) \qquad k = 1, 2, 3, \ldots,$$

$$\xi_k(x) = \xi_k^*(x) - H(x), \quad G_x(z) = P(\xi_k(x) < z) \qquad k = 1, 2, 3, \ldots,$$

$$\xi^{*(n)}(x) = \sum_{k=1}^{n} \xi_k^*(x), \xi^{(n)}(x) = \sum_{k=1}^{n} \xi_k(x), G_x^{(n)}(z) = P(\xi^{(n)}(x) < z) \qquad n = 1, 2, 3, \ldots$$

In order to simplify notations, in the sequel the parameter x of the defined symbols is possibly omitted.

LEMMA 3.1. Suppose that $\xi^{(n)}$ has a symmetrical (to zero) probability distribution for n = 1, 2, 3, ..., and for every $0 < v \le V$ there exists $E(e^{v\xi_k^*})$. Then for arbitrary $0 < v \le V$ and $\varepsilon > 0$ the relation

$$P\left(\left|\frac{\xi^{*(n)}}{n} - H(x)\right| \ge \varepsilon\right) \le 2 \exp\{-n(v(\varepsilon + H(x)) - \ln E[e^{v\xi_k^*}])\} \text{ is valid.} \qquad (3.5)$$

PROOF. By symmetry of the distribution of $\xi^{(n)}$ we have

$$P\left(\left|\frac{\xi^{*(n)}}{n} - H(x)\right| \ge \varepsilon\right) = P(|\xi^{(n)}| \ge n\varepsilon)$$

$$\le P(\xi^{(n)} \ge n\varepsilon) + P(-\xi^{(n)} \ge n\varepsilon) = 2P(\xi^{(n)} \ge n\varepsilon).$$

Hence, for

$$\varepsilon = \frac{t + \ln E[e^{v\xi^{(n)}}]}{nv} > 0 \qquad (t > 0)$$

analogously to (3.4) we obtain (3.5), as

$$P(\xi^{(n)} \ge n\varepsilon) = P(\xi^{(n)} \ge \frac{t+\ln E[e^{v\xi^{(n)}}]}{v}) \le e^{-t} =$$

$$= \exp\{-nv\varepsilon + \ln E[e^{v\xi^{(n)}}]\} = \exp\{-n(v(\varepsilon+H(x))-\ln E[e^{v\xi_k^*}])\}.$$

REMARK 3.1. The symmetry assumption on the distribution of $\xi^{(n)}$ is satisfied in the practically important case when the distribution of $\xi_k^*(x)$, $k = 1,2,3,\ldots$ is symmetrical to $H(x)$. Obviously, instead of that it is sufficient to assume that there holds $P(\xi^{(n)} \ge z) \ge P(-\xi^{(n)} \ge z)$.

In order to sharpen inequality (3.5), one has to maximize (with respect to v) the expression

$$F(v) = v(\varepsilon+H(x))-\ln E[e^{v\xi_k^*}]).$$

Under simple analytical assumptions this can be easily done (for details cf. [15]), hence the following assertion is valid.

THEOREM 3.1. If the value of $H(x)$ is estimated on the base of i.i.d. realizations of the r.v.'s $\xi_k^*(x) = h(x,y_k)$, $k = 1,2,3,\ldots$, then

$$n \ge n_*(H(x),\varepsilon,\delta) = \ln\frac{2}{\delta} \cdot \frac{1}{v_*(\varepsilon+H(x))-\ln E[e^{v_*\xi_k^*(x)}]} \qquad (3.6)$$

implies (3.2); the value v_* is defined by the equation

$$F(v_*) = \max_{0 < v \le V} F(v).$$

REMARK 3.2. The value of n in (3.2) can be estimated also in the (practically relevant) case when v_* is unknown. If e.g. for the given argument x and a suitable $v > 0$ a positive lower bound $C(x,\varepsilon,v)$ for $F(v)$ can be calculated, then (instead of (3.6)) the estimate

$$n \ge \ln\frac{2}{\delta} \cdot \frac{1}{C(x,\varepsilon,v)}$$

can be applied. If e.g. there holds

$$E[e^{v\xi_k(x)}] \le \exp\{\frac{1}{2}B(x)v^2\} \qquad 0 < v \le V_1(x) \qquad (3.7)$$

$(B(x)$, $V_1(x)$ are positive constants depending on x), then it is easy to show that

$$n \geq \ln \frac{2}{\delta} \cdot \frac{2B(x)}{\varepsilon^2} \qquad (0 < \varepsilon \leq B(x) \cdot V_1(x))$$

implies (3.2). Moreover, (3.7) is equivalent to the assumption $E[e^{v\xi_k(x)}] < \infty$ $\quad 0 < v \leq V_2(x)$ ($V_2(x)$ is a constant depending on x), while $B(x)$ can be chosen as any positive number which exceeds $E[\xi_k^2(x)]$.

For illustrating the presented general framework, consider now the special case when the unknown probability $0 < p < 1$ of a random event A is to be estimated on the base of its relative frequency $\frac{\xi^{*(n)}(A)}{n}$ (i.e. $\xi^{*(n)}(A)$ is the number of occurrences of A among n independent trials with the possible outcomes A and $\bar{A}$). In the mentioned case - applying also Remark 3.1. - the described technique reduces to the following special result (cf. [15]).

COROLLARY 3.1. Let $0 < p \leq \frac{1}{2}$, $0 < \varepsilon \leq p(1-p)$. Then

$$P(|\frac{\xi^{*(n)}(A)}{n} - p| \geq \varepsilon) \leq 2\exp\{-nC_*(\varepsilon,p)\},$$

where

$$C_*(\varepsilon,p) = \ln[(\frac{\varepsilon+p}{p})^{\varepsilon+p}(\frac{1-p}{1-p-\varepsilon})^{\varepsilon+p-1}].$$

It is easy to see that

$$C_*(\varepsilon,p) > C(\varepsilon,p) = \frac{\varepsilon^2}{2p(1-p)(1 + \frac{\varepsilon}{2p(1-p)})^2}, \qquad 0 < p \leq \frac{1}{2}, \quad 0 < \varepsilon \leq p(1-p),$$

i.e. Corollary 3.1 improves Bernstein's results. This fact confirms the hope that the suggested modification of the Bernstein-technique may yield improved probability bounds, i.e. makes possible cost (computer time) reduction in the course of solving stochastic optimization problems.

REFERENCES

[1]   Bernstein, S.N.: Probability Theory. (In Russian). Gostechizdat,
       Moscow, 1946.

[2]   Boender, C.G.E., Rinnooy Kan, A.H.G., Timmer, G.T. and Stougie,
       L.: A Stochastic Method for Global Optimization. Math. Pro-
       gramming 22 (1982) 125-140.

[3]   Devroye, L.P.: Progressive Global Random Search of Continuous
       Functions. Math. Programming 15 (1978) 330-342.

[4]   Dorfman, R., Jacoby, H.D. and Thomas, H.A. (eds.): Managing
       Regional Water Quality. Harvard University Press, Cambridge,
       Massachusetts, 1972.

[5]   Ermoliev, Yu. M.: Stochastic Programming Methods (in Russian).
       Nauka, Moscow, 1975.

[6]   Hoeffding, W.: Probability Inequalities for Sums of Bounded
       Random Variables. J. Amer. Stat. Ass. 58 (1963) 13-30.

[7]   Kall, P. and Prékopa, A. (eds.): Recent Results in Stochastic
       Programming. Lecture Notes in Economics and Mathematical
       Systems 179. Springer, Berlin-Heidelberg-New York, 1980.

[8]   Marks, D.H.: Models in Water Resources. In: US EPA: A Guide to
       Models in Governmental Planning and Operations, 103-137.
       EPA, Washington, D.C., 1974.

[9]   Marti, K.: Random Search in Optimization Problems as a Sto-
       chastic Decision Process (Adaptive Random Search). Methods
       of Opns. Res. 36 (1980) 223-234.

[10]   Okamoto, M.: Some Inequalities Relating to the Partial Sum of
       Binomial Probabilities. Ann. Inst. Stat. Math. 10 (1958)
       29-35.

[11]   Pintér, J.: On a Stochastic Model of Reservoir System Sizing.
       In: Proc. 9th IFIP Conf. Opt. Techn. (Iracki, K., Malanowski,
       K., and Walukiewicz, S., eds.), 546-558. Lecture Notes in
       Control and Information Sciences 23. Springer, Berlin-
       Heidelberg-New York, 1980.

[12]   Pintér, J.: On a Random Search Method for Unconstrained Optimi-
       zation (in Russian). Automation and Remote Control 41 (1980)
       n. 12, 76-85.

[13]   Pintér, J.: Stochastic Procedures for Solving Optimization Pro-
          blems (in Hungarian). Alk. Mat. Lapok 7 (1981) 217-252.

[14]   Pintér, J.: Hybrid Procedures for the Solution of Non-Smooth
          Constrained Stochastic Problems (in Russian). Moscow
          University Comp. Math. Cyb. 1982, n. 1, 39-49.

[15]   Pintér , J.: A Modified Bernstein-Technique for Estimating
          Noise-Perturbaded Function Values. Submitted to: Calcolo,
          1983.

[16]   Pintér , J.: Convergence Properties of Stochastic Optimization
          Procedures. To appear in: Math. Operationsforsch. Stat.,
          Ser. Opt. (1984).

[17]   Polyak, B.T.: Nonlinear Programming Methods in the Presence of
          Noise. Math. Programming 14 (1978) 87-97.

[18]   Prékopa, A.: Stochastic Programming Models for Inventory Control
          and Water Storage Problems. In: Colloquia Math. Soc. J.
          Bolyai 7. Inventory Control and Water Storage, (Prékopa,A.,
          ed.) 229-245. North Holland, Amsterdam, 1973.

[19]   Prékopa, A. and Szántai, T.: Flood Control Reservoir System
          Design Using Stochastic Programming. Math. Programming
          Study 9 (1978) 138-151.

[20]   Solis, F.J. and Wets, R.J.-B.: Minimization by Random Search
          Techniques. Math. of Opns. Res. 6 (1981) 19-30.

[21]   Somyódy, L., Herodek, S. and Fischer, J. (eds.): Eutrophication
          of Shallow Lakes: Modeling and Management. The Lake Balaton
          Case Study. IIASA CP-83-S3, Laxenburg, Austria, 1983.

[22]   Törn, A.A.: Global Optimization as a Combination of Global and
          Local Search. Abo Swedish University School of Economics,
          Technical Report A:13. Abo, 1974.

[23]   Yudin, D.B.: Stochastic Programming Problems and Methods (in
          Russian). Sovietskoye Radio, Moscow, 1979.

A METHOD OF FEASIBLE DIRECTIONS FOR SOLVING

NONSMOOTH STOCHASTIC PROGRAMMING PROBLEMS

ANDRZEJ RUSZCZYŃSKI

Instytut Automatyki, Politechnika Warszawska
00-665 Warszawa, Poland

A method of feasible directions is proposed for solving
stochastic programming problems with nonsmooth and nonconvex
objectives. Convergence with probability 1 to stationary
points is proved by means of a special Lyapunov function
technique.

1. INTRODUCTION

Let F be a real-valued function on R^n and let $X \subset R^n$. In this
paper we describe and analyse a stochastic subgradient algorithm
for solving the following problem

minimize F(x)

subject to x ∈ X.
$$\tag{1}$$

We assume that the set X is convex and compact and is known
explicitly, while F may be nonsmooth, nonconvex with its values
and subgradients difficult to calculate. This is typical of
stochastic programming problems with objectives of the form
$F(x) = E f(x,\theta)$, where θ is a random parameter and E denotes
the expected value. In many such problems f is defined by means
of a certain inner optimization problem, which usually causes
its nondifferentiability.

We assume that F is Lipschitz continuous in an open bounded
set X containing X and its subgradient (generalized gradient

sets $\partial F(x)$ possess the following property: there exists a constant σ such that for every $x \in X$ and each $g \in \partial F(x)$ one has

$$F(y) \geq F(x) + \langle g, y-x \rangle - \sigma |y-x|^2 \qquad (2)$$

for all $y \in X$. The family of such functions is identical with the class of functions which are lower-C^1 on X [7,8].

Finally, we assume that at each point x^k generated by the algorithm one can observe, instead of an exact subgradient, a random vector ξ^k such that

$$E \{\xi^k / x^k\} \in \partial F(x^k). \qquad (3)$$

Such vectors (we shall call them stochastic subgradients) can be easily obtained for many classes of stochastic programming problems [3].

Our algorithm uses stochastic subgradients ξ^k to construct a random sequence of points $\{x^k\} \subset X$ convergent with probability 1 to stationary points of (1). To this end an auxiliary filter is used which averages stochastic subgradients observed and produces more stable estimates $\{z^k\}$ which are in turn used in auxiliary direction-finding subproblems. These subproblems consist in minimizing a convex quadratic function in X and reduce to simple quadratic programming problems in case of X being a polyhedron. Feasible directions thus obtained are used to iterate $\{x^k\}$.

Our method is close in spirit to stochastic methods of feasible directions [1,9] elaborated for smooth problems. Its main feature is that it deals with nondifferentiability in a more direct way than another approach based on smoothing the objective by introducing an artificial noise to the method [4,

6]. Our technique for proving convergence, based on a special Lyapunov function, appears to be new, too.

We use $|\cdot|$ to denote the Euclidean norm in R^n. For a sequence $\{x^k\}_{k=0}^{\infty}$ we use K to denote an infinite subset of the set of natural numbers and $\{x^k\}_{k \in K}$ denotes the subsequence associated with K. The subdifferential of F at x is written $\partial F(x)$. We denote by (Ω, B, P) a probability space and use $E\{\xi / B_k\}$ to denote the conditional expectation of a random vector $\xi: \Omega \rightarrow R^n$ with respect to the σ-subfield $B_k \subset B$. We use the abbreviation a.s. for "almost surely".

2. THE METHOD AND ASSUMPTIONS

The method constructs two sequences: main iterates $\{x^k\} \in X$ and ε-subgradient estimates $\{z^k\} \in R^n$. It starts from a feasible x^0 with $z^0=0$ and performs for $k = 0,1,2,\ldots$ the following operations.

Step 1 (direction finding).

$$\text{minimize} \quad [\ \eta(\tilde{x};x^k,z^k) = \langle z^k, \tilde{x}-x^k \rangle + \tfrac{1}{2}|\tilde{x}-x^k|^2\ ]$$
$$\text{subject to} \quad \tilde{x} \in X \tag{4}$$

finding a solution $\tilde{x}^k$, and set

$$d^k = \tilde{x}^k - x^k. \tag{5}$$

Step 2. Choose a stepsize

$$0 < t_k < \min(1,1/a),$$

where a>0 is a certain constant, and set

$$x^{k+1} = x^k + t_k d^k. \tag{6}$$

Step 3. Observe a stochastic subgradient ξ^{k+1} of F at x^{k+1}

and set

$$z^{k+1} = z^k + at_k (\xi^{k+1} - z^k).\qquad(7)$$

Increase k by 1 and go to Step 1.

Let us briefly comment on the method. In the direction finding subproblem one can equivalently define $\tilde{x}^k$ as the orthogonal projection of $x^k - z^k$ onto X, similarly to [6], but we construct z^k from stochastic subgradients at the points x^k, without any artificial random perturbations.

It is also worth noting that we use in the main iteration formula (6) and in the filter (7) the same stepsizes $\{t_k\}$, while in earlier works on methods with gradient averaging it was usually assumed that the main algorithm is infinitely slower than the filter, which significantly simplified the analysis [1,3,4,6,9]. Our approach is close to that of [10].

In the next sections we shall prove that a.s. each accumulation point x´ of the sequence $\{x^k\}$ generated by (4)-(7) is stationary for (1), i.e. there exists $g \in \partial F(x´)$ such that

$$\langle g, x-x´\rangle \geq 0 \quad \text{for all } x \in X.$$

In case of a convex objective this condition is also sufficient.

We shall now formulate assumptions which will be used in our analysis.

Let B_k be the minimal σ-subfield which measures $\{x^0, z^0, \dots, x^{k-1}, z^{k-1}, x^k\}$ and let E_k denote the conditional expectation with respect to B_k. We make the following assumptions.

(H1) $E_k \xi^k \in \partial F(x^k)$ for k = 1, 2, ...

(H2)   There is s > 0 such that $E_k |\xi^k|^2 \leq s$ a.s.

for k = 1, 2, ...

(H3) For each $k \geq 1$ the stepsize t_k is B_{k+1}-measurable.

(H4) $\sum_{k=1}^{\infty} E \, t_k^2 < \infty$.

(H5) $\sum_{k=1}^{\infty} t_k = \infty$ a.s. .

Conditions (H1)-(H5) are typical of recursive stochastic optimization algorithms (cf., e.g., [3,5]). Let us observe,
however, that in (H3) we allow dependence of t_k on both x^k
and z^k, i.e. on d^k, which may be useful in practice.

3. THE LYAPUNOV FUNCTION

In this section we define a special continuous function
$W(x,z)$ and prove that our method (4)-(7) is in a certain
stochastic sense a descent method for this function.

Let $\hat{n}(x,z)$ denote the optimal value of the direction finding subproblem (4), i.e.

$$\hat{n}(x,z) = \min_{\tilde{x} \in X} [\langle z, \tilde{x}-x \rangle + \tfrac{1}{2}|\tilde{x}-x|^2].$$
(8)

Let us define the function $W: X \times R^n \rightarrow R$ as follows

$$W(x,z) = a \, F(x) - \hat{n}(x,z).$$
(9)

In what follows we shall estimate the difference $W(x^{k+1}, z^{k+1})$
$- W(x^k, z^k)$. We shall use the following notation:

$$\xi^k = g^k + r^k, \quad g^k \in \partial F(x^k), \quad E_k r^k = 0;$$
(10)

$$z^k = z_g^k + z_r^k,$$

$$z_g^{k+1} = z_g^k + at_k (g^{k+1} - z_g^k), \quad z_g^0 = g^0 \in \partial F(x^0),$$
(11)

$$z_r^{k+1} = z_r^k + at_k (r^{k+1} - z_r^k), \quad z_r^0 = - z_g^0.$$
(12)

We start from the following simple observation.

LEMMA 1. The sequence $\{z_g^k\}$ is bounded and $\{z_r^k\} \rightarrow 0$ a.s. .

Proof. Since X is compact, the subgradients $\{g^k\}$ are uniform-

ly bounded [2,7], which proves the first assertion. To prove
the second one, observe that (12) is a simple stochastic app-
roximation algorithm for minimizing $\phi(z_r) = \frac{1}{2}a|z_r|^2$, which is
convergent under (H1)-(H5) (see, e.g., [3,5]).

LEMMA 2. The following statements are true:

(a) $\sum_{k=0}^{\infty} |x^{k+1} - x^k|^2 < \infty$ a.s.,

(b) $\sum_{k=0}^{\infty} |z^{k+1} - z^k|^2 < \infty$ a.s.,

(c) the series $\sum_{k=0}^{\infty} t_k \langle d^k, r^{k+1} \rangle$ is convergent a.s. .

Proof. Assertion (a) follows immediately from (H4) and the
compactness of X, since $E|x^{k+1} - x^k|^2 = E\{t_k^2|d^k|^2\} \leq C E t_k^2$,
where C is the diameter of X. To prove (b) observe that

$$|z^{k+1} - z^k|^2 = a^2 t_k^2 |\xi^{k+1} - z^k|^2 \leq 2a^2 t_k^2 (|\xi^{k+1}|^2 + |z^k|^2).$$

Owing to (H2)-(H4), one has $\sum_{k=0}^{\infty} t_k^2 |\xi^{k+1}|^2 < \infty$ a.s., and
$\{z^k\}$ is bounded by Lemma 1. Finally, from (H1) and (H3) we
get $E_{k+1}\{t_k \langle d^k, r^{k+1} \rangle \} = 0$, i.e. the series $\sum_{k=0}^{\infty} t_k \langle d^k, r^{k+1} \rangle$
is a martingale, convergent by virtue of (H2)-(H4) and the
boundedness of $\{d^k\}$. The proof is complete.

We are now ready to prove the main result of this section.

LEMMA 3. There exists a sequence of random variables $\{u_k\}$
such that for k = 0, 1, 2, ...

$$W(x^{k+1}, z^{k+1}) - W(x^k, z^k) \leq - at_k|d^k|^2 + u_k \qquad (13)$$

and

$$\sum_{k=0}^{\infty} u_k < \infty \quad \text{a.s. .} \qquad (14)$$

Proof. Consider at first the function ñ defined by (8). Ob-
serve that $\langle z, \tilde{x} - x \rangle + \frac{1}{2}|\tilde{x} - x|^2 = \frac{1}{2}|\tilde{x} - x + z|^2 - \frac{1}{2}|z|^2$, which im-
plies that the solution $\tilde{x}(x,z)$ to (8) is the orthogonal pro-

jection of x-z onto the convex set X. Thus $\tilde{x}(x,z)$ is unique for all x and z, and $\hat{\eta}$ is differentiable at each point (x,z) (see, e.g., [2]) with the gradient

$$\nabla_x \hat{\eta}(x,z) = -z + x - \tilde{x}(x,z),$$

$$\nabla_z \hat{\eta}(x,z) = \tilde{x}(x,z) - x.$$

Since $\tilde{x}(x,z)$, as the orthogonal projection, is Lipschitz continuous, these gradients are Lipschitz continuous in (x,z), too. Hence a constant C exists such that for each k

$$- \hat{\eta}(x^{k+1}, z^{k+1}) + \hat{\eta}(x^k, z^k) \le \langle z^k - x^k + \tilde{x}^k, x^{k+1} - x^k \rangle +$$
$$\langle x^k - \tilde{x}^k, z^{k+1} - z^k \rangle + C(|x^{k+1} - x^k|^2 + |z^{k+1} - z^k|^2).$$

Using (5)-(7) we obtain

$$- \hat{\eta}(x^{k+1}, z^{k+1}) + \hat{\eta}(x^k, z^k) \le t_k \langle z^k, d^k \rangle + t_k |d^k|^2 -$$
$$at_k \langle d^k, \xi^{k+1} \rangle + at_k \langle z^k, d^k \rangle +$$
$$C(|x^{k+1} - x^k|^2 + |z^{k+1} - z^k|^2). \tag{15}$$

Since $\tilde{x}^k$ is the projection of $x^k - z^k$ onto X, one must have

$$\langle x^k - z^k - \tilde{x}^k, x^k - \tilde{x}^k \rangle \le 0,$$

which yields the inequality

$$\langle z^k, d^k \rangle \le - |d^k|^2.$$

Combining this with (15) one gets

$$- \hat{\eta}(x^{k+1}, z^{k+1}) + \hat{\eta}(x^k, z^k) \le - at_k |d^k|^2 - at_k \langle d^k, g^{k+1} \rangle -$$
$$- at_k \langle d^k, r^{k+1} \rangle + C(|x^{k+1} - x^k|^2 + |z^{k+1} - z^k|^2). \tag{16}$$

Let us now use (2) with $y = x^k$ and $x = x^{k+1}$:

$$F(x^{k+1}) - F(x^k) \le \langle g^{k+1}, x^{k+1} - x^k \rangle + \sigma |x^{k+1} - x^k|^2 =$$
$$t_k \langle g^{k+1}, d^k \rangle + \sigma |x^{k+1} - x^k|^2. \tag{17}$$

Adding (17) multiplied by the constant a to (16) we obtain

the required inequality

$$W(x^{k+1}, z^{k+1}) - W(x^k, z^k) \leq - at_k |d^k|^2 + u_k,$$

where

$$u_k = - at_k \langle d^k, r^{k+1} \rangle + (C + \sigma) |x^{k+1} - x^k|^2 + C |z^{k+1} - z^k|^2.$$

Relation (14) results from Lemma 2. The proof is complete.

4. PROPERTIES OF ε-SUBGRADIENT ESTIMATES

In this section we investigate properties of the sequence $\{z^k\}$ generated by the method (4)-(7). Let us at first review basic properties of ε-subdifferentials of functions satisfying (2).

We define the ε-subdifferential of F at a point $x \in X$ with $\varepsilon \geq 0$ as follows

$$\partial_\varepsilon F(x) = \{g \in R^n : F(y) \geq F(x) + \langle g, y - x \rangle - \sigma |y - x|^2 - \varepsilon$$

$$\text{for all } y \in X\}.$$

The elements of the set $\partial_\varepsilon F(x)$ will be called ε-subgradients of F at x. Our definition of $\partial_\varepsilon F(x)$ does not essentially differ from that for convex functions, except for the residual term $\sigma |y - x|^2$. Therefore one can easily derive the following properties of $\partial_\varepsilon F$:

(a) for each $\varepsilon \geq 0$ and all $x \in X$ the set $\partial_\varepsilon F(x)$ is convex and compact;

(b) if $g^1 \in \partial_{\varepsilon_1} F(x)$ and $g^2 \in \partial_{\varepsilon_2} F(x)$, $x \in X$, $\varepsilon_1 \geq 0$ and $\varepsilon_2 \geq 0$, then $\lambda g^1 + (1-\lambda) g^2 \in \partial_{\lambda \varepsilon_1 + (1-\lambda) \varepsilon_2} F(x)$;

(c) the mapping $(x, \varepsilon) \rightarrow \partial_\varepsilon F(x)$ is upper-semicontinuous and bounded, i.e. if $x^k \rightarrow x \in X$, $\varepsilon_k \geq 0$, $\varepsilon_k \rightarrow \varepsilon$, $g^k \in \partial_{\varepsilon_k} F(x^k)$, then the sequence $\{g^k\}$ is bounded and all its accumulation points belong to $\partial_\varepsilon F(x)$.

Proof of these properties is almost identical with that for the convex case and will be omitted (see, e.g. [7]).

LEMMA 4. <u>There exist nonnegative random variables</u> ε_k, $k = 0$, 1, 2, ..., <u>such that for all k one has</u>

$$z_g^k \in \partial_{\varepsilon_k} F(x^k) \tag{18}$$

<u>and</u>

$$\lim_{k \to \infty} \varepsilon_k = 0 \quad \text{a.s.} \tag{19}$$

<u>Proof.</u> By definition, $z_g^0 \in \partial F(x^0)$, i.e. $\varepsilon_0 = 0$. Suppose that for some k there is $\varepsilon_k \geq 0$ such that $z_g^k \in \partial_{\varepsilon_k} F(x^k)$. Then for all $y \in X$

$$F(y) \geq F(x^k) + \langle z_g^k, y - x^k \rangle - \sigma|y - x^k|^2 - \varepsilon_k =$$
$$F(x^{k+1}) + \langle z_g^k, y - x^{k+1} \rangle - \sigma|y - x^{k+1}|^2 -$$
$$[\varepsilon_k + F(x^{k+1}) - F(x^k) - \langle z_g^k, x^{k+1} - x^k \rangle +$$
$$2\sigma\langle y - x^k, x^{k+1} - x^k \rangle].$$

The function F is Lipschitz continuous, the set X is bounded and the sequence $\{z_g^k\}$ is bounded by Lemma 1. Hence for all y one may write the inequality

$$F(y) \geq F(x^{k+1}) + \langle z_g^k, y - x^{k+1} \rangle - \sigma|y - x^{k+1}|^2 - \bar{\varepsilon}_k$$

where

$$\bar{\varepsilon}_k = \varepsilon_k + C|x^{k+1} - x^k|$$

and C is a certain constant. Thus $z_g^k \in \partial_{\bar{\varepsilon}_k} F(x^{k+1})$. Since $g^{k+1} \in \partial_0 F(x^{k+1})$ and $0 < at_k < 1$, using property (b) one obtains from (11) the relations

$$z_g^{k+1} \in \partial_{\varepsilon_{k+1}} F(x^{k+1}),$$ (20)

$$\varepsilon_{k+1} = (1 - at_k)\, \bar{\varepsilon}_k = (1 - at_k)(\varepsilon_k + C|x^{k+1} - x^k|).$$ (21)

This proves (18) for all k. It remains to prove (19). From (21) we get

$$0 \le \varepsilon_{k+1} \le \varepsilon_k - at_k\varepsilon_k + Ct_k|d^k|.$$

Let $\delta > 0$. Obviously, $|d^k| \le |d^k|^2/\delta + \delta$. Thus for all k

$$0 \le \varepsilon_{k+1} \le \varepsilon_k - t_k(a\varepsilon_k - C\delta) + (C/\delta)t_k|d^k|^2.$$ (22)

Let us use Lemma 3. The function W is bounded from below on $X \times R^n$ since F is continuous on X and $\hat{h}(x,z) \le 0$ for all $x \in X$ and $z \in R^n$. Therefore Lemma 3 implies that

$$\Sigma_{k=0}^{\infty}\, t_k|d^k|^2 < \infty \quad \text{a.s. .}$$ (23)

The right side of (22) describes one iteration of the Robbins--Monro method for solving the equation $a\varepsilon = C\delta$ and thus (23) combined with (H3)-(H5) yields

$$\limsup_{k \to \infty} \varepsilon_k \le C\delta/a \quad \text{a.s. .}$$

Since $\delta > 0$ is arbitrary, the proof of (19) is complete.

We can now easily prove the following result.

LEMMA 5. <u>With probability 1 each accumulation point</u> $(x^{\check{}}, z^{\check{}})$ <u>of the sequence</u> $\{(x^k, z^k)\}$ <u>satisfies the relation</u>

$$z^{\check{}} \in \partial F(x^{\check{}}).$$

<u>Proof.</u> Let $\{(x^k, z^k)\}_{k \in K} \to (x^{\check{}}, z^{\check{}})$. By Lemma 4, $z_g^k \in \partial_{\varepsilon_k} F(x^k)$ and $\varepsilon_k \to 0$ a.s. . Then property (c) of the ε-subdifferential implies that each accumulation point of $\{z_g^k\}_{k \in K}$ is in $\partial F(x^{\check{}})$. But $\{z_r^k\} \to 0$ a.s. by Lemma 1, hence $\{z_g^k\}_{k \in K} \to z^{\check{}} \in \partial F(x^{\check{}})$, as required.

5. CONVERGENCE

Having derived basic properties of our Lyapunov function (Lemma 3) and ε-subgradient estimates (Lemma 5), we are now ready to prove convergence of our method.

THEOREM. Assume (H1)-(H5). Then a.s. each accumulation point $(x\check{},z\check{})$ of the sequence $\{(x^k,z^k)\}$ generated by the method (4)-(7) satisfies the relations

(a) $x\check{}$ is stationary for (1);

(b) $z\check{} \in \partial F(x\check{})$.

Additionally, the sequence $\{F(x^k)\}$ is convergent a.s. .

Proof. From (23) and (H4) we get

$$\liminf_{k \to \infty} |d^k| = 0. \tag{24}$$

Since $\hat{\mathsf{n}}(x,z) \leq 0$ for each $x \in X$ and $\hat{\mathsf{n}}$ is continuous, (24) and Lemma 1 imply that

$$\limsup_{k \to \infty} \hat{\mathsf{n}}(x^k,z^k) = 0. \tag{25}$$

Suppose that one can find $\alpha > 0$ and an infinite set of indices K such that $\hat{\mathsf{n}}(x^k,z^k) \leq -\alpha$ for $k \in K$. Then there exists $\varepsilon > 0$ such that if $|x^j - x^k| + |z^j - z^k| \leq \varepsilon$ then $\hat{\mathsf{n}}(x^j,z^j) \leq -\alpha/2$. Since $|\hat{\mathsf{n}}(x^j,z^j)| \leq (|z^j| + \frac{1}{2}|d^j|)|d^j|$ and both $\{z^j\}$ and $\{d^j\}$ are bounded, one can find $\beta > 0$ such that $|d^j|^2 > \beta$ whenever $|x^j - x^k| + |z^j - z^k| \leq \varepsilon$, $k \in K$.

Let us define for each $k \in K$ the index

$$l(k) = \min \{l \geq k: |x^l - x^k| + |z^l - z^k| > \varepsilon\}.$$

By hypothesis, $l(k) < \infty$, since otherwise one would obtain a contradiction with (25). For each $k \in K$ Lemma 3 yields the inequality

$$W(x^{1(k)}, z^{1(k)}) \leq W(x^k, z^k) - a\Sigma_{j=k}^{1(k)-1} t_j |d^j|^2 + \Sigma_{j=k}^{1(k)-1} u_j$$

$$\leq W(x^k, z^k) - a\beta\Sigma_{j=k}^{1(k)-1} t_j + \Sigma_{j=k}^{1(k)-1} u_j. \quad (26)$$

By the definition of $1(k)$,

$$\varepsilon < |x^{1(k)} - x^k| + |z^{1(k)} - z^k| \leq \Sigma_{j=k}^{1(k)-1} t_j |d^j| +$$

$$a\Sigma_{j=k}^{1(k)-1} t_j |g^{j+1} - z^j| + a|\Sigma_{j=k}^{1(k)-1} t_j r^{j+1}|.$$

Since the sequences $\{d^j\}$, $\{g^j\}$ and $\{z^j\}$ are bounded, there

exists a constant C such that for all $k \in K$ one has

$$\varepsilon \leq C\Sigma_{j=k}^{1(k)-1} t_j + a|\Sigma_{j=k}^{1(k)-1} t_j r^{j+1}|.$$

This combined with (26) yields

$$W(x^{1(k)}, z^{1(k)}) - W(x^k, z^k) \leq - a\beta\varepsilon/C +$$

$$(a^2\beta/C)|\Sigma_{j=k}^{1(k)-1} t_j r^{j+1}| + \Sigma_{j=k}^{1(k)-1} u_j. \quad (27)$$

Let $k \to \infty$, $k \in K$. By (H1)-(H4) the series $\Sigma_{j=0}^{\infty} t_j r^{j+1}$ is a

convergent martingale and thus

$$\lim_{k \to \infty} |\Sigma_{j=k}^{1(k)-1} t_j r^{j+1}| = 0.$$

By Lemma 3,

$$\lim_{k \to \infty} \Sigma_{j=k}^{1(k)-1} u_j = 0.$$

Therefore the right side of (27) tends to $- a\beta\varepsilon/C$ as $k \to \infty$,

$k \in K$. On the other hand, it follows from Lemma 3, that the

sequence

$$s_k = W(x^k, z^k) + \Sigma_{j=k}^{\infty} u_j, \quad k = 1, 2, \ldots$$

is nonincreasing and bounded a.s., hence convergent a.s..

Thus $\{W(x^k, z^k)\}$ is convergent a.s. and the left side of

(26) tends to 0. We obtain a contradiction, which proves

that

$$\lim_{k \to \infty} \hat{\eta}(x^k, z^k) = 0. \tag{28}$$

From (28) by the continuity of $\hat{\eta}$ we immediately get that

$$\hat{\eta}(x^{\check{}}, z^{\check{}}) = 0 \tag{29}$$

at any accumulation point $(x^{\check{}}, z^{\check{}})$. Since $x^{\check{}} \in X$ by construction and $z^{\check{}} \in \partial F(x^{\check{}})$ by Lemma 5, (29) is equivalent to the stationarity of $x^{\check{}}$. Additionally, convergence a.s. of $\{W(x^k, z^k)\}$ and (28) imply convergence a.s. of $\{F(x^k)\}$. The proof is complete.

It is worth noting that the method not only finds stationary points but also subgradients that appear in necessary conditions of optimality.

REFERENCES

[1] L. G. Bazhenov and A. M. Gupal, On a certain analogue of the method of feasible directions, *Kibernetika (Kiev)*, 1973, no. 9, pp. 94-95 [Russian].

[2] F. H. Clarke, *Optimization and nonsmooth analysis*, Wiley, New York, 1983.

[3] Yu. M. Ermoliev, *Stochastic programming methods*, Nauka, Moscow, 1976 [Russian].

[4] A. M. Gupal, A method for minimization of functions satisfying the Lipschitz condition, *Kibernetika (Kiev)*, 1980, no. 2, pp. 91-94 [Russian].

[5] H. J. Kushner and D. S. Clark, *Stochastic approximation methods for constrained and unconstrained systems*, Springer, New York, 1978.

[6] F. Mirzoakhmedov and M. V. Mikhalevich, A method with projection of stochastic quasi-gradients, *Kibernetika (Kiev)*, 1983, no. 4, pp. 103-109 [Russian].

[7] E. A. Nurminski, *Numerical methods for solving deterministic and stochastic minimax problems*, Naukova Dumka,
Kiev, 1979 [Russian].

[8] R. T. Rockafellar, Favorable classes of Lipschitz continuous functions in subgradient optimization, in: *Progress in nondifferentiable optimization*, E. A. Nurminski (ed.), IIASA, Laxenburg, 1982, pp. 125-144.

[9] A. Ruszczyński, Feasible direction methods for stochastic programming problems, *Mathematical Programming*,
19 (1980), pp. 220-229.

[10] A. Ruszczyński and W. Syski, Stochastic approximation
algorithm with gradient averaging for unconstrained
problems, *IEEE Transactions on Automatic Control*,
AC-28 (1983), pp. 1097-1105.

A PROBABILISTIC ANALYSIS OF THE SET PACKING PROBLEM

C. VERCELLIS
Dipartimento di Matematica - Univ. di Milano

Abstract

The maximum cardinality Set Packing Problem (SPP), formulated as $\max\{e_n x: Ax \underset{=}{\le} e_m, \ x \epsilon \{0,1\}^n\}$, where A is a mxn binary matrix and e_n, e_m are vectors of 1's of appropriate size, is a well-known NP-hard integer programming problem arising in a number of real-word applications.

In this paper, a probabilistic analysis of the SPP is developed, considering a stochastic model of the incidence matrix A in which the entries are independent Bernoulli distributed random variables, each with probability p_n of being equal to 1. A threshold function $t_k(n,p_n)$ on the number m of constraints is derived for the property that a packing of cardinality k exists with probability tending to one as n tends to infinity.

In particular, it is shown that, if the probability p_n is constant, then the optimum value is almost surely equal to 1; thus, combining the latter result with the corresponding one for the SCP obtained in [4], the duality gap is analyzed and shown to be asymptotically large as log n in ratio almost surely.

Finally, in § 4, the performance of the simple "blind" sequential algorithm is investigated, and a sufficient condition is assigned which the sequences p_n and m_n have to satisfy for the ratio of the optimal solution value to the approximate one to be asymptotically bounded by 2 almost surely.

1. Introduction

Let E be a finite set of m elements, which can be represented, without loss of generality, as the set of integers $\{1,2,\ldots m\}$. Let also $G=\{F_j \subseteq E, j \in N\}$, with $N=\{1,2,\ldots n\}$, be a collection of subsets of E.

A subcollection $G' \subseteq G$ is said to be a *packing* if the condition : $F_i \cap F_t = \emptyset$ holds for every pair of distinct subsets F_i, $F_t \in G'$. A *k-packing* is defined to be a packing which contains exactly k subsets.

In the *Set Packing Problem* (SPP), it is required to determine a packing of maximum cardinality in G. The SPP is an *NP-hard* problem in combinatorial optimization, for which an ε-approximation algorithm is unlikely to exist, for any $\varepsilon>0$ (see $[2]$, p.146). In the case in which $|F_j|=2$ for all subsets $F_j \in G$, the SPP reduces to the problem of finding a maximum independent set of vertices in a graph, still remaining NP-hard.

The SPP can be formulated as an integer programming problem, associating to each subcollection $G' \subseteq G$ a characteristic vector $x \in \{0,1\}^n$, and defining a mxn incidence matrix $A= [a_{ij}]$, $i \in E$, $j \in N$, such that : $a_{ij}=1$ if $i \in F_j$, and $a_{ij}=0$ otherwise. Thus, the SPP can be stated as

$$
\begin{aligned}
\max \quad & e_n x \\
\text{s.to} \quad & Ax \le e_m \\
& x \in \{0,1\}^n,
\end{aligned}
\tag{1}
$$

where e_n, e_m are vectors of 1's of size, respectively, n and m.

Related to the SPP is the *Set Covering Problem* (SCP), formulated as follows: given a set E and a collection G as above, find a *cover*, i.e. a subcollection $G^o \subseteq G$ such that

$$
\bigcup_{F_j \in G^o} F_j = E,
$$

which is of minimum cardinality. It can be easily recognized that the SCP admits a $\{0,1\}$ programming formulation :

$$\min \quad e_n x$$

$$\text{s.to} \quad Ax \geq e_m$$

$$x \in \{0,1\}^n .$$

It turns out that SPP and SCP are a pair of dual integer problems, in the sense that the LP-relaxation of the SPP in (1) and that of the SCP whose incidence matrix is given by A^T, represent a pair of LP-dual problems. Denoting as z^p and z^c, respectively, the optimum values associated to the SPP in (1) and to the SCP

$$\min\{e_m x: A x \geq e_n, \ x \in \{0,1\}^m\}, \tag{2}$$

it follows by the theory of duality that $z^p \leq z^c$. The analysis of the difference $d = z^c - z^p$, termed *duality gap*, appears of interest because several "additive" implicit enumeration algorithms for solving the SPP (or the SCP) consider the optimum of the dual integer for upper (or lower) bounding the optimum of the subproblem at the given node of the search tree.

In this paper, a probabilistic analysis of the SPP is developed, considering a stochastic model of the incidence matrix A, described in Section 2, in which the entries are independent Bernoulli distributed random variables, each with probability $p = p_n$ of being equal to 1.

In Section 3, a threshold function $t_k(n,p)$ on the number $m = m_n$ of constraints is derived for the property that a packing of cardinality k exists with probability tending to one as n tends to infinity. It is also shown that, if the probability p_n is constant, then the optimum value of the SPP is almost surely equal to 1; combining this result with the corresponding one for the SCP obtained in [4], the duality gap is shown to be asymptotically large as log n in ratio.

Finally, in Section 4, the performance of the simple *blind sequential algorithm* is analysed, and a sufficient condition is assigned which the sequences m_n and p_n have to satisfy for the ratio of the optimum value to the approximate one to be asymptotically bounded by 2 almost surely. The blind algorithm has been analysed also in [1] and [3] for the particular case of the maximum independent set in a graph.

2. The stochastic model

A *Random Set Packing Problem* (RSPP) is characterized by a triple (n,m,p) where n,m are integers and $p \in (0,1)$, and is defined to be a SPP with m elements, n subsets and a $m \times n$ incidence matrix A whose entries are independent identically distributed random variables (r.v.), each with a Bernoulli distribution of parameter p, i.e.

$$Pr\{a_{ij}=1\} = p = 1-q, \quad i \in E, \ j \in N.$$

The number n of subsets will be considered the leading parameter in the sequel, in the sense that the behavior of certain sequences of r.v.'s (the sequence of optima, or that of approximate values generated by the blind algorithm), defined on a RSPP of parameters (n,m,p) will be analysed in terms of stochastic convergence as n grows asymptotically large and both $m=m_n$, $p=p_n$ satisfy given conditions as functions of n. All the sequences $\{m_n\}$ considered throughout the paper are assumed to be monotonically increasing and polinomially bounded in n, i.e. $m/n^{\alpha} \to 0$ for $n \to \infty$, for some $\alpha>0$.

The following notations will be useful in the sequel. Let $\Omega_k=\{S \subseteq G: |S| = k\}$ be the set of the $\binom{n}{k}$ subcollections of G containing exactly k subsets. Define, over the set 2^G of all subcollections $G' \subseteq G$, an indicator function $\delta(.)$, such that :

$$\delta(G') = \begin{cases} 1 & \text{if } G' \text{ is a packing,} \\ 0 & \text{otherwise.} \end{cases}$$

Let $T_k=\{S \in \Omega_k : \delta(S)=1\}$ be the set of all k-packings, and $Y_k=|T_k|$ be their number. Of course, with respect to the RSPP defined above, both T_k and Y_k are r.v.'s; to make this fact explicit the notations $T_k(n,m,p)$ and $Y_k(n,m,p)$ will be sometime preferred.

The optimum value of a RSPP of parameters (n,m,p) will be denoted as

$$B_n=B(n,m,p) = \max\{k: Y_k(n,m,p)>0\}.$$

Finally, the standard notations $\lceil x \rceil$ and $\lfloor x \rfloor$ will be used to indicate respectively the least integer not less than x and the greatest integer not greater than x.

3. A threshold on the number of elements

In this section, a threshold $t_k(n,p)$ on the number of elements m is derived, for the property that the RSPP does not contain a k-packing.

Let $P_k(n,m,p)=\Pr\{Y_k(n,m,p) \geq 1\}$ be the probability that the RSPP contains a k-packing. Then, $t_k(n,p)$ is said to be a *threshold* on m for the property $\{Y_k(n,m,p) \geq 1\}$ if the two following conditions hold:

(i) $\quad \lim_{n \to \infty} \dfrac{t_k(n,p)}{m} = 0 \quad \Rightarrow \quad \lim_{n \to \infty} P_k(n,m,p)=0;$

(ii) $\quad \lim_{n \to \infty} \dfrac{t_k(n,p)}{m} = \infty \quad \Rightarrow \quad \lim_{n \to \infty} P_k(n,m,p) = 1.$

As a preliminary result, we have the following lemma:

Lemma 3.1: Let $S_o=\{1,2,\dots k\}$; then, for $k \geq 2$,

$$\Pr\{\delta(S_o) = 1\} = \left[q^k + kpq^{k-1} \right]^m.$$

Proof : $\Pr\{\delta(S_o)=1\} = \Pr\{\forall i \in E: \sum_{j=1}^{k} a_{ij} \leq 1\} =$

$$= \left[\Pr\{ \sum_{j=1}^{k} a_{ij} \leq 1 \} \right]^m = \left[q^k + kpq^{k-1} \right]^m \quad \blacktriangle$$

The following lemma gives bounds from below and from above on $P_k(n,m,p)$:

Lemma 3.2: For any integer $k \geq 2$:

$$1- \left\{1- \left[q^k+kpq^{k-1} \right]^m \right\} ^{\left\lfloor \frac{n}{k} \right\rfloor} \quad \leq P_k(n,m,p) \leq \binom{n}{k} \left[q^k+kpq^{k-1} \right]^m.$$

Proof: We first derive the upper bound:

$$P_k(n,m,p) = \Pr\{Y_k(n,m,p) \geq 1\} =$$

$$= \Pr\{ \exists\, S \varepsilon \Omega_k : \quad \delta(S) = 1\}$$

$$\leq \binom{n}{k} \Pr\{ \delta(S_o) = 1\}$$

$$= \binom{n}{k} \left[q^k + kpq^{k-1} \right]$$

by lemma 3.1, and the bound from above is established.

In order to derive the lower bound, observe that it is possible to choose a collection of $\lfloor \frac{n}{k} \rfloor$ k-subsets $S_j \varepsilon \Omega_k$, $j=1,2,\ldots \lfloor \frac{n}{k} \rfloor$, such that the S'_j s are pairwise disjoint. Therefore we have

$$P_k(n,m,p) = \Pr\{Y_k(n,m,p) \geq 1\}$$

$$= 1 - \Pr\{Y_k(n,m,p) = 0\}$$

$$= 1 - \Pr\{\forall S \varepsilon \Omega_k : \delta(S) = 0\}$$

$$\geq 1 - \Pr\{\delta(S_j) = 0, \quad j=1,2,\ldots \lfloor \frac{n}{k} \rfloor \}$$

$$= 1 - \left[1 - \Pr\{\delta(S_o)=1\} \right]^{\lfloor \frac{n}{k} \rfloor}$$

$$= 1 - \{ 1 - \left[q^k + kpq^{k-1} \right]^m \}^{\lfloor \frac{n}{k} \rfloor} \quad ,$$

which is the required result. ▲

Theorem 3.1: The threshold function for the property $\{Y_k(n,m,p) \geq 1\}$ is given, for $k \geq 2$, by :

$$t_k(n,p) = \frac{\log n}{\log\left[q^k + kpq^{k-1} \right]^{-1}}$$

Proof : Let $z_k(p) = \left[q^k + kpq^{k-1} \right]^{-1}$; suppose first that

$$m = \left\lceil \frac{\omega_n \log n}{\log z_k(p)} \right\rceil$$

for some sequence $\{\omega_n\}$ such that: $\lim_{n \to \infty} \omega_n = \infty$.

Then, using lemma 3.2, we have

$$P_k(n,m,p) \leqslant \binom{n}{k} \left[q^k + kpq^{k-1} \right]^m$$

$$\leqslant \frac{1}{k!} \exp\{k\log n - m\log z_k(p)\}$$

$$\leqslant \frac{1}{k!} \exp\{(k-\omega_n) \log n\} \to 0$$

as $n \to \infty$, showing that condition (i) is satisfied.

Suppose now that $m = \left\lfloor \frac{\varepsilon_n \log n}{\log z_k(p)} \right\rfloor$ for some sequence $\{\varepsilon_n\}$ such that :

$\lim_{n \to \infty} \varepsilon_n = 0$. Then, again by lemma 3.2,

$$P_k(n,m,p) \geqslant 1-\{1- \left[q^k + kpq^{k-1} \right]^m \}^{\left\lfloor \frac{n}{k} \right\rfloor}$$

$$\geqslant 1-\{1- n^{-\varepsilon_n}\}^{\left\lfloor \frac{n}{k} \right\rfloor}$$

$$= 1-\exp\{ \left\lfloor \frac{n}{k} \right\rfloor \log(1-n^{-\varepsilon_n})\} \to 1$$

as $n \to \infty$, so that also condition (ii) is satisfied. ▲

We turn now the attention to the case in which the probability p of success in the RSPP is constant, independent of n. In this case the threshold in theorem 3.1 reduces to $t_k(n,p) = \log n$; we have therefore the

Corollary 3.1: For constant p, and $k \geqslant 2$:

(a) if $\dfrac{m}{\log n} \to 0$ then $P_k(n,m,p) \to 1$ as $n \to \infty$;

(b) if $\dfrac{m}{\log n} \to \infty$ then $P_k(n,m,p) \to 0$ as $n \to \infty$,

and $\lim\limits_{n \to \infty} B_n = 1$ in pr.

As already remarked in Section 1, it is of interest to analyse the duality gap between the optimum of the SPP and that of the dual SCP defined in (2). In particular, the asymptotic behaviour of the doubly indexed sequence $\{D_{n,m}\}$ (where $D_{n,m}$ is the r.v. representing the duality gap for a (n,m,p) RSPP) can be investigated in terms of stochastic convergence as $n \to \infty$ and m satisfies given conditions as a function of n. In $\begin{bmatrix} 4 \end{bmatrix}$ the model of random SCP corresponding to the RSPP has been considered; in particular it has been shown that:

Theorem 3.2: For constant p, and for a random SCP with n elements and m

subsets:

(a) if $\lim\limits_{n \to \infty} \dfrac{m}{\log n} = 0$ then the SCP is a.e.

infeasible;

(b) if $\lim\limits_{n \to \infty} \dfrac{m}{\log n} = \infty$ then $\lim\limits_{n \to \infty} \dfrac{C_n}{\log n} = \dfrac{1}{\log(1-p)^{-1}}$ a.e.,

where C_n denotes the optimum of the SCP.

In light of part (a) of the latter theorem, the only significant case for analysing the behaviour of $\{D_{n,m}\}$ is that in which $\dfrac{m}{\log n} \to \infty$ as $n \to \infty$. Indeed, combining the parts (b) of corollary 3.1 and theorem 3.2, we have the

Corollary 3.2: For constant p, if $\lim\limits_{n \to \infty} \dfrac{m}{\log n} = \infty$

then $\lim\limits_{n \to \infty} \dfrac{D_{n,m}}{\log n} = \dfrac{1}{\log(1-p)^{-1}}$ in pr.

The latter result says that, assuming the stochastic model of SPP and SCP described in Section 2 (which has also been considered in the literature

to generate random instances of the two problems for testing exact algorithms),
the duality gap grows to infinity as fast as $\Theta(\log n)$, where n is the number
of variables of the SPP - as well as the number of constraints of the SCP.

4. Analysis of the "blind" algorithm

The most simple approximation algorithm for solving the SPP is based on
a sequential "blind" strategy: the list of subsets $\{F_1, F_2, \ldots F_n\}$ is
scanned, and the subsets are selected according to the rule "retain a subset
iff it does not intersect any of the previously retained subsets".

The blind algorithm B can be stated formally as follows:

procedure B;
    input : collection $G = \{F_1, F_2, \ldots F_n\}$;
    output : packing $T \subseteq G$;
    begin $T := \emptyset$;
       for i=1 to n do
          if (for each $F_j \in T : F_i \cap F_j = \emptyset$) then
            $T: = T \cup \{F_i\}$;
       end;
end.

Obviously, when algorithm B halts, the set T represents a maximal
packing (i.e. a packing not contained properly in any other packing), so that
its cardinality can be retained as an approximation to the unknown optimum
of the SPP.

Let $R_n = R_n(m,p)$ be the cardinality of the set T generated by the B
algorithm on a (n,m,p) instance of the RSPP.

It will be shown that the conditions

C1 : $\lim_{n \to \infty}$ $p \log n = 0$;

C2 : $\lim_{n \to \infty}$ $mp^2 = \alpha$ for some $\alpha \in (0,1)$;

are sufficient for guaranteeing that the ratio B_n/R_n of the optimum value B_n to the approximate value R_n is almost surely bounded by 2 as n tends to infinity. Observe that conditions C1 and C2 are naturally satisfied in most significant cases. For instance, in $\begin{bmatrix}3\end{bmatrix}$ it has been shown that the ratio B_n/R_n is bounded by 2 a.e. for the independent set problem on a model of random graph which is essentially equivalent to a particular case of the RSPP with $p=\dfrac{2}{n}$ and $m=p'\binom{n}{2}$, with $p' \in (0,1)$: it is easy to see that both conditions C1 and C2 are satisfied in this case. The same is true when the RSPP models h-uniform hypergraphs, with $p=\dfrac{h}{n}$ and $m= p'\binom{n}{h}$, $p' \in (0,1)$.

A bound from above on B_n is first established by means of the

Theorem 4.1: Under conditions C1 and C2, the sequence of r.v.'s B_n satisfies

$$\limsup_{n\to\infty} \frac{B_n}{\log n} \leq \frac{2}{\alpha} \quad \text{a.e.}$$

Proof: Let $k=k(n) = \left\lceil (1+\varepsilon)\dfrac{2 \log n}{\alpha} \right\rceil$, for any $\varepsilon>0$; then, by lemma 3.2 and Stirling's formula, one has

$$\Pr\{ B_n \geq k\} \leq \binom{n}{k} \left[q^k + kpq^{k-1} \right]^m$$

$$\leq \frac{n^k}{k!} \left[q^k + kpq^{k-1} \right]^m$$

$$= \frac{n^k}{k!} \exp\{m \log \left[q^{k-1}(1+(k-1)p) \right] \}$$

$$= \frac{n}{k!} \exp\{ (k-1)\log n+m \left[(k-1)\log(1-p) +\log(1+(k-1)p) \right] \} .$$

By condition C1 it follows that $\lim_{n\to\infty} kp = 0$, so that

$$\Pr\{B_n \geq k\} \leq \frac{n}{k!} \exp\{ (k-1)\left[\log n - k\frac{mp^2}{2} + o(kpm^2) \right] \} .$$

Moreover, by condition C2, one has that

$$\lim_{n \to \infty} \quad \log n - k \frac{mp^2}{2} + o(kmp^2) =$$

$$= \lim_{n \to \infty} \quad - \varepsilon \log n + o(\log n) = -\infty .$$

Thus, for large n, it follows

$$Pr\{B_n \geqslant k\} \leqslant \frac{n}{k!} = o(n^{-i})$$

for any integer i>0, so that the series $\sum_n Pr\{B_n \geqslant k\}$ is convergent. Therefore, Borel-Cantelli lemma ensures that

$$\limsup_{n \to \infty} \quad \frac{B_n}{k(n)} \leqslant 1 \quad \text{a.e.}$$

Letting k(n) be expressed as

$$k(n) = \frac{2 \log n}{\alpha} \quad (1+\varepsilon) + \Theta_n, \quad 0 \leqslant \Theta_n < 1,$$

one can easily see that

$$\limsup_{n \to \infty} \quad \frac{B_n}{\log n} \leqslant \frac{2}{\alpha} \quad \text{a.e.} \quad \blacktriangle$$

The second step towards analysing the asymptotic behaviour of the ratio B_n/R_n is taken in the

Theorem 4.2: Under conditions C1 and C2, the sequence $\{R_n\}$ satisfies

$$\liminf_{n \to \infty} \quad \frac{R_n}{\log n} \geqslant \frac{1}{\alpha} \quad \text{a.e.}$$

Proof: Suppose that the algorithm B is applied to an infinite sequence of subsets randomly generated according to the stochastic model of Section 2.

Specifically, for each n let $\{F_j^{(n)}\}$, j=1,2,...n,..... be an infinite sequence of subsets of E={1,2,...m}, such that any element e$\in$E has probability p=p_n to belong to any subset $F_j^{(n)}$, independently of any other element or subset. Hence, the first n subsets of the sequence $\{F_j^{(n)}\}$ represent an instance (n,m,p) of the RSPP.

For each n, suppose that the algorithm B is applied to the sequence $\{F_j^{(n)}\}$, and let $H_i^{(n)}$ be the r.v. expressing the number of subsets retained by B among $\{F_1^{(n)}, \ldots F_2^{(n)}, \ldots F_i^{(n)}\}$ - i.e. $H_i^{(n)}$ represents the cardinality of the partial solution T at the i-th iteration.

Let $\{Q_j^{(n)}\}$ be a sequence of r.v.'s defined as

$$Q_0^{(n)} = 0 \quad , \quad Q_j^{(n)} = \min \{i: H_i^{(n)} = j\} ,$$

and representing, for each j, the index of the j-th selected subset. Let also $W_j^{(n)} = Q_{j+1}^{(n)} - Q_j^{(n)}$, $j=0,1,2,\ldots$, indicate the number of iterations in which the partial solution contains exactly j subsets; clearly, we have

$$Q_j^{(n)} = \sum_{i=0}^{j-1} W_i^{(n)} .$$

Let

$$U_j^{(n)} = \bigcup_{i=1}^{j} F_i^{(n)}$$

be the set of elements of E belonging to the first j subsets included in T.

For the r.v.'s of the sequence $\{W_j^{(n)}\}$ we have

$$r_j^{(n)} = \Pr\{W_j^{(n)} > t\} = \sum_{s=0}^{m} (1-q^s)^t \Pr\{ |U_j^{(n)}| = s\} \qquad (3).$$

Moreover, $r_j^{(n)} \leq r_{j+1}^{(n)}$, $j=0,1,\ldots$.

Let $k = k(n) = \left\lceil (1-\varepsilon) \dfrac{\log n}{\alpha} \right\rceil$, for any $\varepsilon > 0$. Then,

$$\Pr\{R_n < k\} = \Pr\{H_n^{(n)} < k\} = \Pr\{Q_k^{(n)} > n\}$$

$$\leq \Pr\{W_j^{(n)} > \left\lceil \frac{n}{k} \right\rceil , \text{ some } j=0,1,\ldots k-1\}$$

$$\leq \sum_{j=0}^{k-1} \Pr\{W_j^{(n)} > \left\lceil \frac{n}{k} \right\rceil \}$$

$$\leq k \Pr\{W_k^{(n)} > \left\lceil \frac{n}{k} \right\rceil \}$$

$$\leq k(1-q^{\lceil kmp \rceil})^{\left\lceil \frac{n}{k} \right\rceil} + k \Pr\{ |U_k^{(n)}| > \lceil kmp \rceil \} , \qquad (4)$$

where the last inequality derives from (3). As far as the first term in (4) is concerned, we have

$$k(1-q^{\lceil kmp \rceil})^{\left\lceil \frac{n}{k} \right\rceil} \leq k \exp\{ \frac{n}{k} \log (1-q^{\lceil kmp \rceil}) \}$$

$$\leq k \exp\{ - \frac{n}{k} \exp\{ \lceil kmp \rceil \log q \}$$

$$= k \exp\{ - \frac{n}{k} \exp\{ -kmp^2 + o(kmp^2) \}$$

$$= k \exp\{ - \frac{n^{\varepsilon+o(1)}}{k} \}, \qquad (5)$$

where the last equality comes from condition C2.

Consider now the second term in (4) and observe that $p+q/k \sim 1/k$:

$$\Pr\{ |U_k^{(n)}| > \lceil kmp \rceil \} \leq km \binom{n}{k} (1-q^k-kpq^{k-1})^{kmp}$$

$$\leq m \frac{n^k}{(k-1)!} (1-q^k-kpq^{k-1})^{kmp}$$

$$= \frac{m}{(k-1)!} \exp\{k \log n + kmp \log(1-q^k-kpq^{k-1})\}$$

$$\leq \frac{m}{(k-1)!} \exp\{k(\log n - mp)\} \leq \frac{m}{(k-1)!} . \qquad (6)$$

Combining (4), (5) and (6), and recalling that m is polynomially bounded in n, it follows that the series $\sum_n \Pr\{R_n < k\}$ is convergent, and the Borel-Cantelli lemma leads to the required result. ▲

As an immediate consequence of the Theorems 4.1 and 4.2, one deduces the

Corollary 4.1: Under conditions C1 and C2, the sequences of r.v.'s $\{B_n\}$ and $\{R_n\}$ satisfy

$$\limsup_{n\to\infty} \frac{B_n}{R_n} \leq 2 \quad \text{a.e.}$$

References

[1] G. Ausiello, A. Marchetti-Spaccamela, M. Protasi, "Probabilistic analysis of the performance of greedy strategies over some combinatorial problems", R. 21, IASI-CNR 1981.

[2] M.R. Garey, D.S. Johnson, "Computers and intractability: a guide to the theory of NP-completeness, Freeman, San Francisco, 1979.

[3] G.R. Grimmett, C.J.H. McDiarmid, "On coloring random graphs", Math. Proc. Camb. Phil. Soc. 77 (1975).

[4] C. Vercellis, "A probabilistic analysis of the set covering problem", Annals of Operations Research 1 (1984).

Lecture Notes in Control and Information Sciences

Edited by M. Thoma

Lecture Notes in Control and Information Sciences

Edited by M. Thoma

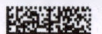